东方风情
会所/餐饮细部解析

Oriental Flavor
DETAILED ANALYSIS OF CLUB & RESTAURANT

深圳视界文化传播有限公司 编

大连理工大学出版社

图书在版编目(CIP)数据

东方风情:会所/餐饮细部解析:汉英对照/深圳视界文化传播有限公司编. —大连:大连理工大学出版社, 2013.6
　　ISBN 978-7-5611-7769-3

Ⅰ. ①东… Ⅱ. ①深… Ⅲ. ①休闲娱乐–服务建筑室内装饰设计–作品集–世界–现代②餐馆–室内装饰设计–作品集–世界–现代 Ⅳ. ①TU247

中国版本图书馆CIP数据核字(2013)第068768号

出版发行：大连理工大学出版社
　　　　　（地址：大连市软件园路80号 邮编：116023）
印　　刷：深圳市彩美印刷有限公司
幅面尺寸：230mm×300mm
印　　张：18
出版时间：2013年6月第1版
印刷时间：2013年6月第1次印刷
责任编辑：裘美倩
责任校对：王丹丹
装帧设计：（香港）视界国际出版有限公司

ISBN 978-7-5611-7769-3
定　　价：288.00元

电话：0411-84708842
传真：0411-84701466
邮购：0411-84703636
E-mail: designbooks_dutp@yahoo.com.cn
URL: http://www.dutp.cn

如有质量问题请联系出版中心：（0411）84709246　84709043

PREFACE 前言

Oriental style is full of elegance and connotation, civility and richness. There are various kinds of flavors, but it lies the most in the rhyme of orient. "*Oriental Flavor*" consists of space display, design interpretation and detailed description to allow readers comprehensively appreciate characteristic space designs with oriental culture. Those cases listed in the book are all the latest high-grade oriental style designs, and the pictures of all cases are beautiful and clear, each with oriental features. The explanation of design concept is detailed with complete plans as well as part of the elevation and detail design drawings of such things as material, lighting, wall, floor, ceiling, door and window, interior landscape and so on, which is with high reference sense.

东方之清雅含蓄、端庄丰华，风情有百种，最在韵之东方。《东方风情》在于空间展示+设计解读+细部解构，让读者全方位领略具有东方文化的特色空间设计。书中撷取之案例皆为最新的具有高品位的东方风格设计典范，所有项目之图片皆精美清晰，各具东方特色，设计理念解说详细，配有完整的平面图，以及部分项目材料、照明、墙体、地面、天花、门窗、室内景观等的立面及细部设计图，可参考性极强。

CONTENTS 目录

006
GU YI GE — SPACE ZEN WITH "NO LIKE NO MATCH"
古逸阁——无象无相的空间禅意

016
ENJOYING THE COUNTRYSIDE LIFE WITH WINE
杯酒话山居

030
YUE FU CLUB
悦府会

046
HAILIANHUI — DESCRIBING THE FASHIONABLE NEW ORIENT
海联汇——写意时尚新东方

056
THE ORCHID PAVILION NO.1
兰亭壹号

066
LIU RESIDENCE
刘家大院

072
YING SHENG BRAND — A FLASH OF COLLISION
盈盛号——碰撞出来的光芒

080
HUA JIAN TANG — THE SEASON YARD IN ZHOU ZHUANG
花间堂·周庄季香院

092
IMPRESSION OF FEET — MASSAGE
印象足道

102
LUCKY AND PEACEFUL FOR A LIFETIME
祥和百年

114
IMPRESSION OF HAKKA
印象客家

130
NATIONAL YARD
国民院子

136
SUZHOU GARDEN NO.1
苏园壹号

144
JIULIHE RESTAURANT
九里河餐厅

152
BEIJING LONGTAN LAKE PARAMOUNT CHAMBER
龙潭湖九五书院

162
IMPRESSION AT JIANGNAN
RESTAURANT — WANDA SHOP
印象望江南餐厅万达店

172
MANGO THAI RESTAURANT,
NINGBO
宁波美泰泰国餐厅

180
ORIENTAL COURTYARD
东方大院

188
HAO ZI ZAI TEAHOUSE
好自在茶艺馆

196
FOUR SEASONS MIN FU ROAST
DUCK RESTAURANT
四季民福烤鸭店

204
YILAN COURTYARD
HEALTHY CLUB
意兰庭保健会所

210
TI XIANG YI TEAHOUSE
提香溢茶楼

218
CAI DIE XUAN IN
YANGZHOU
扬州采蝶轩

226
ORIGINAL MEAL
原膳

234
JINYU RESTAURANT IN
FOSHAN
佛山锦裕食府

242
HOKKAIDO TAPPASAKI
北海道铁板烧

254
UNDERSTAND WORLD TEA
观茶天下

268
POMEGRANATE
BLOSSOM HALL
榴花溪堂

280
PIN YI SALON
品奕造型

GU YI GE — SPACE ZEN WITH "NO LIKE NO MATCH"

古逸阁——无象无相的空间禅意

The spatial form of tea club is often restricted in certain kind of ideology, but the difference of the space with the same name is reflected in people's imaginative creation to its character. As one of the designer's "Floating Cloud" series of works, Gu Yi Ge tea club's "no like no match" theme is a new interpretation of its space atmosphere. The driving source of this creation makes the club derive a different kind of temperament, making the wood as the intermediary and Zen in mind while the calm heart helps business and the poor life calms the heart.

Design Company 设计公司
Fuzhou Damu Heshi Design 福州大木和石设计联合会馆
Designer 设计师
Jie Chen 陈杰
Project Location 项目地点
Fuzhou, Fujian 福建福州
Project Area 项目面积
185 m²
Photographer 摄影师
Yuedong Zhou 周跃东
Text Editing 文字
Jianqing Quan 全剑清
Main Materials 主要材料
Old wood, flax cloth, acrylic panel, rice paper, gray mirror, floating cloud lamp, etc.
旧木、麻布、亚克力板、宣纸、灰镜、浮云灯等

茶会所的空间形态往往被束缚在某种意识形态中，但相同名称的空间种类的差异性体现在人们对其个性的匠心独运。古逸阁茶会所作为设计师"浮云"系列作品之一，"无象无相"是其对空间意境的崭新诠释。这种创造的源动力让这个会所衍生出别样的气质，以木为媒，以禅为念，心清助道业，清苦得心定。

古逸阁茶会所位于浦上大道，与万达商圈毗邻。虽地处繁华闹市，但设计师遵循"物尽其用是为俭"的理念，将一份古朴与清静浸润在空间之中，让目之所及的一切愈加耐人寻味。会所前的户外区域，地面用夯实的枕木铺陈，周边的桌椅以木质、石质、竹质交糅在一起，透着一股自然苍劲的美，悄然打动着过往的人们。墙面的透明玻璃呈现出会所内部的景致，它仿佛是一个取景框，涵盖的风景或许是一个插着枯枝的陶罐，一把改良过后的中式椅子，抑或是灯光留下的影子，骤然生动。

引导人们进入会所内部的地板是从附近老房子拆迁得来的旧木，凹凸不平的纹理自成风景，而那些或深或浅的不同色泽仿佛在默守一段尘封的往事，留给人遐想的空间，同时也衬托出这屋子的素雅氛围。内部空间的墙面也适时地使用到了这些旧木，它们带着一股时过境迁的淡淡忧伤，但当我们留恋起儿时的记忆和味道，时光就此倒转，骨子里的文化归属让设计更加赏心悦目，让我们更容易找到共鸣。此外，会所内的柜体、台面、搁物架、门窗均以旧木作为设计的载体。或许它们已破旧、已衰败，但设计师挖掘出它们所蕴藏的蓬勃生命力，并将之运用在茶会所中，这无疑也使他心中对传统文化的理解落到了实处。

除了旧木饰墙，天然麻布也是空间中重要的装裱材料，素而不俗。前台区域的顶上悬挂着若干浮云状照明灯具，它们在光影的烘托下和谐地律动，并实现了空间氛围的蜕变。前台区域的背后是一个包厢，古朴自然的材质在其间和谐共处着。这些物件模糊了时间的概念，然颇有一番自在的个性。在与临近功能区域的衔接上，"无象无相"是设计师追求的意境。一面灰砖砌成的墙，以方格作为透视，并点缀上烛光，让诗意在这古朴的空间中悠悠不尽，没有了浮躁，不见了仓皇。一旁的走道用青石做踏步，周边配以水景与地灯。走道的尽头是一面由石砾组成的墙，下面的石墩上放置着若干松果，寓意菩提。其上方用白色枯枝装饰，在灯光的映衬下显得张力十足。与常规的佛像布置不同，设计师在此区域以意代形，将禅意与生俱来的气质展现得自然天成。走道尽头的左侧是一个独立的品茶区，旧木、老物件依然成为这里的主角，与传统文化一脉相承的灵动也愈加风姿卓然。右侧是一个展示区，茶品、建盏、紫砂壶等物件陈列其中，它们是一种关于古朴与情境的东西，留下的是经过沉淀的生活。

人若谢物，物未必不知。以物为善而无贵贱新旧之分，这是种人生的选择，也是种设计的态度。善用旧物成为设计师解决问题的一种态度，衍生出一种新的生活方式。于是在古逸阁的空间设计中，材质之间的呼应与衬托、线条之间的交织与平衡、几何形态之间的构成与对比没有多余的一笔，不带丝毫拖沓。在这样没有喧哗的交流中，整个空间似乎变得香醇，人们的心情也就明朗了起来。

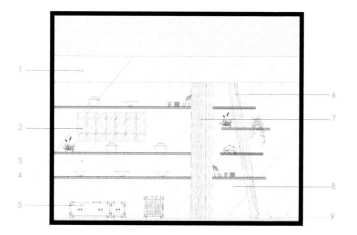

1. White painted and old-processed wall
2. Original window of the building
3. White painted and old-processed wall
4. Wood panel
5. Finished decoration
6. Old wooden peg
7. Old wooden panel
8. White painted and old-processed wall
9. Skirting lines

1. 墙面刷白做旧
2. 原建筑窗户
3. 墙面刷白做旧
4. 木作层板
5. 成品摆件
6. 旧木桩
7. 旧木板
8. 墙面刷白做旧
9. 踢脚线

Facade C of tea area
泡茶区立面图C

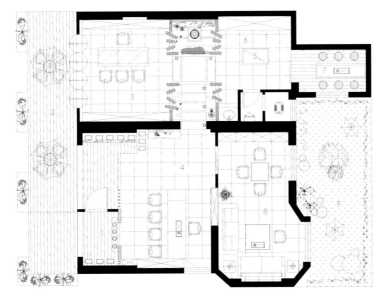

1. Front hall
2. Outdoor leisure area
3. Chat area
4. Main tea area
5. Cabinet
6. Showroom
7. Japanese style tea area
8. Private room
9. Outdoor landscape

1. 前厅
2. 户外休闲区
3. 洽谈区
4. 主泡茶区
5. 中岛柜
6. 展厅
7. 日式泡茶区
8. 包厢
9. 户外造景

Plan layout
平面布置图

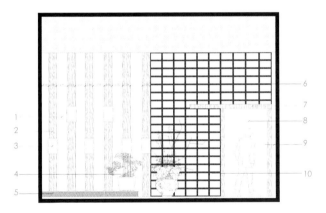

Facade B of tea area
泡茶区立面B

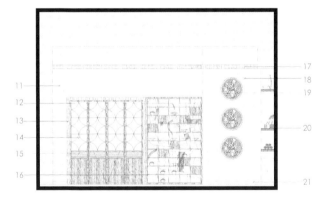

Facade A of tea area
泡茶区立面A

1. White painted and old-processed wall
2. Old wood
3. Embroidery on iron surface
4. Landscape
5. Low lift 150, old wood
6. Prefavricated cement shape wall
7. Old wood decorative door frame
8. Passage
9. White painted and old-processed wall
10. Landscape
11. White painted and old-processed wall

1. 墙面刷白做旧
2. 旧木
3. 铁皮制绣
4. 端景
5. 低抬150、旧木
6. 水泥预制造型墙
7. 旧木装饰门框
8. 通道
9. 墙面刷白做旧
10. 端景
11. 墙面刷白做旧

12. Painting
13. Old wood door frame
14. Rice paper
15. Custom grillwork door
16. Wooden brushing dark brown
17. Old wood
18. White painted and old-processed wall
19. Round plaque
20. Wooded panel
21. Skirting lines

12. 喷绘
13. 旧木门框
14. 宣纸
15. 定制花格门
16. 木作刷深棕色
17. 旧木
18. 墙面刷白做旧
19. 圆圈
20. 木作层板
21. 踢脚线

ENJOYING THE COUNTRYSIDE LIFE WITH WINE
杯酒话山居

This residence located among the green mountain and water was built in the 1970s, the main structure of which is constructed with the strip-style stone. It once was a CS field war base, and is remodeled with new investment, which aims to build a leisure site combining outdoor activity and health preserving as a whole, so that people coming here can contact closely with nature, be close to the blue sky and enjoy the green. After positioning the purpose of the design this time, designers' design work will begin in turn. Designers primarily make some necessary change on the original structure of the building, which are mainly in two aspects: one is the reconstruction of the floors, another is the reconstruction of the building's external environment.

Design Company 设计公司
COMEBER 宽北设计机构

Designer 设计师
Na Xu 许娜

Project Location 项目地点
Fuzhou, Fujian 福建福州

Project Area 项目面积
800 m²

Cost of hard decoration 硬装造价
2.2 million 220万

Main Materials 主要材料
Log, preservative wood, metal bricks, blue bricks, gray tiles, etc.
原木、防腐木、金属砖、青砖、灰瓦片等

掩映于青山碧波间的这一建筑建造于20世纪70年代，其主体结构为条形原石搭砌而成，之前为一CS野战基地，此次投资重新改建，旨在打造一个以户外、养生为一体的休闲场所，使到这里的人可以与自然最近距离地接触，亲近蔚蓝，享受碧绿。定位好此番设计的目的后，设计师的设计工作便依次有序地展开。先是对建筑原结构进行一些必要的改造，主要是两个方面：一个是楼层的改造，一个是建筑外围环境的改造。

原楼顶加盖的一层以青砖灰瓦表现出悠远灵动的东方意境；大面积落地窗能够最大限度地把自然带进室内；依地势引入山泉而成的池塘波光潋滟；不远处依山而建的风雨亭在取材和工艺上传承了中国传统建筑元素，在做旧手法的处理后一眼看过去颇有几分似杜甫草堂。通过改造将自然山水与人文建筑更完美地融合在了一起。

本案在材料的运用上也是颇费了几番心思，为了遵循原建筑的整体风格，也为了与自然环境的沟通融合达到一致，在用材时，木材占了很大的比重，从门窗到桌椅到顶楼大面积木地面再到栏杆等等很多地方都是木料，值得一提的是，这其中很多是从各地拆掉的老房子中淘来的，于是现在可以从很多地方看到岁月留下的烙印。

在陈设上，设计师也是强调旧物利用的原则。很多颇具民风的老式家具也是从多个地方归置来的，把这些旧物置于这样一个全新的空间中，东方风格的秀气典雅得到了新的定义，新旧之间可以更好地契合，更为这一建筑空间增添了几分神韵。

闲暇之余，登上位于半山腰的这里，放眼望去，满山的绿顿时消解了全身的疲惫，再斟上一壶薄酒，此情此景岂不让人快哉！

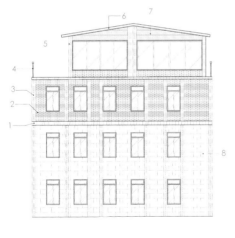

Facade 2
主外观二

Facade 1
主外观一

1. Preservative wood with dark color
2. Old blue bricks with gray seam
3. Blue bricks with black seam
4. Iron railings
5. Blue bricks with black seam
6. Preservative wood with dark color
7. Wooden shutter with wood color painting
8. Stone exterior wall
9. Iron railings
10. Wooden shutter with wood color painting
11. Preservative wood with dark color
12. Blue bricks with black seam
13. Blue bricks with black seam
14. Old preservative wood with dark color
15. Old preservative wood with dark color
16. Preservative wood with dark color
17. Stone exterior wall
18. Stone exterior wall
19. Blue brick hook black seam
20. Dark wooden panels
21. Iron railings
22. Preservative wood with dark color

1. 防腐木擦深色
2. 旧青砖勾灰色缝
3. 青砖勾黑缝
4. 铁艺栏杆
5. 青砖勾黑缝
6. 防腐木擦深色
7. 木制百叶木色油漆
8. 原条石外墙
9. 铁艺栏杆
10. 木制百叶木色油漆
11. 防腐木擦深色
12. 青砖勾黑缝
13. 青砖勾黑缝
14. 旧青砖勾灰色缝
15. 旧青砖勾灰色缝
16. 防腐木擦深色
17. 原条石外墙
18. 原条石外墙
19. 青砖黑缝
20. 深色木条斜拼
21. 铁艺栏杆
22. 防腐木擦深色

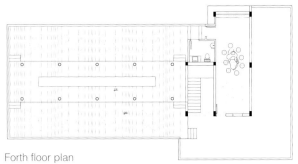

Forth floor plan
四层平面布置图

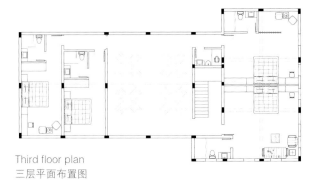

Third floor plan
三层平面布置图

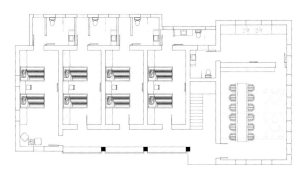

Second floor plan
二层平面布置图

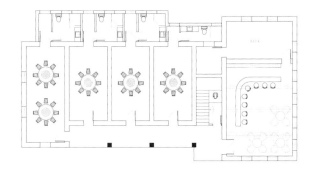

First floor plan
一层平面布置图

YUE FU CLUB
悦府会

This project is based on Park Hyatt Hotel and is close to the natural scenic spot of Dongqianhu, Ningbo. It enjoys exclusively the cultural landscape resources such as the small Putuo and the Southern Song stone inscription, whose geographical location is unparalleled. In space and visual language, it perfectly joints Park Hyatt; in space, the space sequence of Chinese architectural tradition is used to strengthen the Oriental sense of etiquette and honor; in visual sense, exquisite materials and distinctive craft details, with the simple match of black and white, show the flavor of misty rain as well as the ink and wash of Dongqianhu. In hardware and intelligent system, it adheres to the consistent high quality of Park Hyatt Hotel, which inadvertently allows customers to feel the essential Park Hyatt character.

Design Company 设计公司
HARIZON SPACE 深圳市昊泽空间设计有限公司
Designer 设计师
Song Han 韩松
Project Location 项目地点
Ningbo, Zhejiang 浙江宁波
Project Area 项目面积
850 m²
Main Materials 主要材料
White sand beige, Thailand teak, etc. 白沙米黄、虎檀尼斯、泰柚等

1. Porch	1. 门廊
2. Foyer	2. 门厅
3. Sandbox area	3. 沙盘区
4. Reception desk	4. 接待台
5. Model desk	5. 模型台
6. Porch	6. 门廊
7. Light well	7. 采光井
8. Small model area	8. 小模型区
9. Self-service water bar	9. 自助式水吧台
10. Outdoor porch	10. 外廊
11. VIP room 1	11. VIP洽谈室1
12. VIP room 2	12. VIP洽谈室2
13. Operating room	13. 操作间
14. Mop pool	14. 拖把池
15. Light well	15. 采光井
16. Female room	16. 女卫
17. Male room	17. 男卫

First floor plan
一层平面布置图

本项目以柏悦酒店为依托，傍依宁波东钱湖自然景区，独享小普陀、南宋石刻群等人文景观资源，地理位置无可比拟。在空间和视觉语言上与柏悦酒店完美对接；在空间上以中国建筑传统的空间序列强化东方式的礼仪感和尊贵感；在视觉上通过考究的材料和独具匠心的工艺细节，以简约的黑白搭配一气呵成，展现了东钱湖烟雨蒙蒙、水墨沁染的气韵。在硬件和智能化体系上坚持柏悦酒店一贯高品质的传承，让客户不经意间感受到骨子里的柏悦性格。

  设置独立专属的高端客户接待空间，独立酒水吧、独立卫生间。尽享尊贵、专属的接待服务。细分功能空间，将一个空间的多重功能拆解细分，每个都尽善极致，大大提升品质感。

  增加全新的功能体验，在商业行为中加入文化和艺术气质。我们在地下一层设计了一座小型私人收藏博物馆，涉猎瓷器、家具、中国现代绘画、玉器等……不仅大大提升品质，同时也给客户带来视觉和心理上的全新震撼体验。

  身处其中，恍若超脱凡尘，烦恼、杂念消失无踪。带出一抹我独我乐的欢喜。

  正所谓：别业居幽处，到来生隐心。

  南山当户牖，沣水映园林。

  屋覆经冬雪，庭昏未夕阴。

  寥寥人镜外，闲坐听春禽。

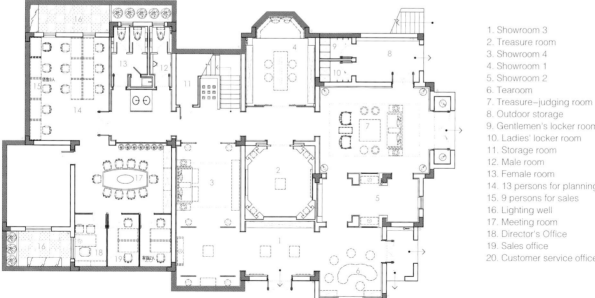

First floor plan
一层平面布置图

1. Showroom 3
2. Treasure room
3. Showroom 4
4. Showroom 1
5. Showroom 2
6. Tearoom
7. Treasure-judging room
8. Outdoor storage
9. Gentlemen's locker room
10. Ladies' locker room
11. Storage room
12. Male room
13. Female room
14. 13 persons for planning
15. 9 persons for sales
16. Lighting well
17. Meeting room
18. Director's Office
19. Sales office
20. Customer service office

1. 展厅3
2. 珍品室
3. 展厅4
4. 展厅1
5. 展厅2
6. 茶室
7. 鉴宝阁
8. 室外储藏室
9. 男更衣室
10. 女更衣室
11. 储藏室
12. 男洗手间
13. 女洗手间
14. 策划13人
15. 销售9人
16. 采光井
17. 会议室
18. 总监办公室
19. 销售办公室
20. 客服办公室

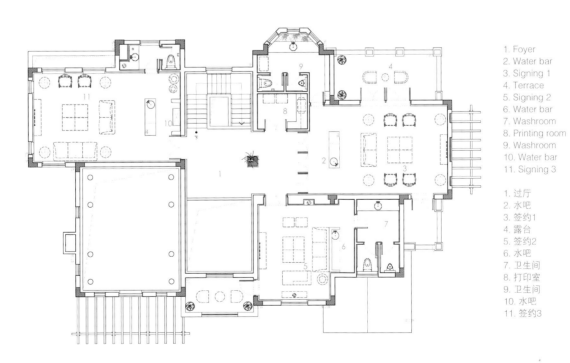

Second floor plan
二层平面布置图

1. Foyer
2. Water bar
3. Signing 1
4. Terrace
5. Signing 2
6. Water bar
7. Washroom
8. Printing room
9. Washroom
10. Water bar
11. Signing 3

1. 过厅
2. 水吧
3. 签约1
4. 露台
5. 签约2
6. 水吧
7. 卫生间
8. 打印室
9. 卫生间
10. 水吧
11. 签约3

HAI LIAN HUI — DESCRIBING THE FASHIONABLE NEW ORIENT

海联汇——写意时尚新东方

HAI LIAN HUI is a newly renovated restaurant, presenting mix-match New Oriental style. As a supporting project of Hailian Hotel, the owner hopes the renovation of the restaurant not only follows the warm color of original hotel's lobby space, but also doesn't lack the mood of oriental culture. Based on this, the designer mix-matches the elements of modern sense and oriental mood. The designer conveys the generous and profound oriental mood through the interwaving of materials, the combination of spatial blocks, the imagery simulation of shape and the changeable match of soft furnishings, which is full of current aesthetic forms. Different from the gorgeous and retro sense of New Oriental space in a general sense, HAI LIAN HUI more emphasizes furnishings, allocation and the creation of cultural atmosphere in commercial space, which focuses on controlling the taste of space and does not lose Oriental mood while simplifying the traditional Chinese elements.

Design Company 设计公司
Fuzhou Chuangyi Weilai Design Co., Ltd.　福州创意未来装饰设计有限公司

Designer 设计师
Yanghui Zheng　郑杨辉

Project Location 项目地点
Fuzhou, Fujian　福建福州

Project Area 项目面积
320 m²

Photographer 摄影师
Yuedong Zhou　周跃东

Cost of Hard Decoration 工程造价
About 1.2 million　约120万

Main Materials 主要材料
Wood, PVC antique wood floor, leather hard decoration, square steel, cane products, etc.
实木、PVC仿古木地板、皮革硬包、防锈处理方钢、藤制品等

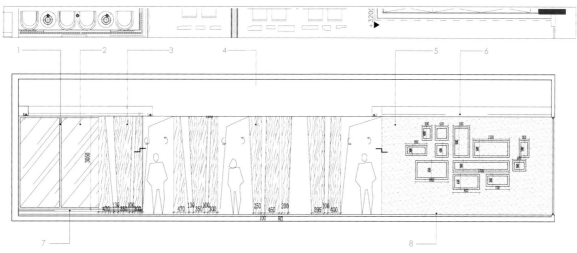

Elevation of car seat area
卡座区立面图

1. 50X50mm rust-proof processing
2. Steel glass
3. Screen
4. Solid wood screen partition
5. Hand-painted wave toxture
6. Built-in light belt
7. Built-in light belt
8. 50 ϕ spotlight

1. 50mmx50mm防锈处理方钢
2. 钢化玻璃
3. 屏风
4. 实木屏风隔断
5. 手刮波浪纹理
6. 内藏灯带
7. 内藏灯带
8. ϕ50射灯

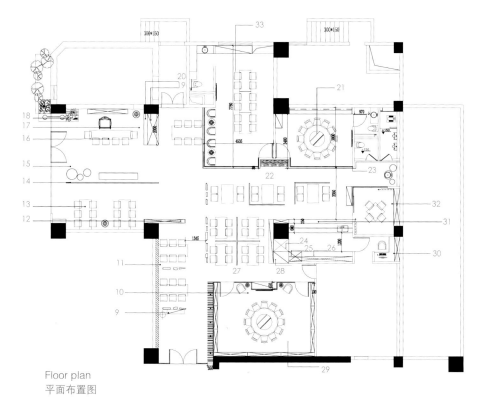

Floor plan
平面布置图

9. Landscape furnishing area	9. 景观陈设区
10. Scene waiting area	10. 情景等候区
11. Deck area 2	11. 卡座区2
12. Book club	12. 书吧
13. 80,90 living space	13. 80、90活态空间
14. Abstract water concept Screen	14. 抽象水概念屏风
15. Waiting area	15. 等候区
16. Tea area	16. 茗茶区
17. Showroom and reception area	17. 展示接待区
18. Scene exhibition space	18. 情景展示空间
19. Goods storage cabinets	19. 餐厅物品储藏柜
20. Lightbox of cultural topics	20. 文化主题灯箱
21. Chinese style private room	21. 中式包间
22. Deck area 1	22. 卡座区1
23. Square steel fixed 20mm from wall	23. 方钢固定离墙20mm
24. High storage cabinet	24. 储藏高柜
25. Coffee machine and other equipments	25. 咖啡机等设备
26. Operating table	26. 操作台
27. Deck area 3	27. 卡座区3
28. Cabinet of restaurant	28. 餐厅杂物柜
29. Theme room	29. 主题包间
30. Office area of kitchen	30. 厨房办公区域
31. Health concept showroom	31. 养身概念陈设展示
32. Chess and card lounge	32. 棋牌、休闲室
33. Western style room	33. 西式包间

　　海联汇是一间重新翻修，呈现混搭新东方风格的餐厅。作为海联酒店的配套项目，业主希望餐厅的重装既要沿用原酒店大堂空间的暖色调，又不乏东方文化的意境，基于此，设计师将现代感与东方意境的元素混搭，在材质的穿插、空间体块的组合、造型的意象模拟和软装陈设的百变拼盘中传达了大气深邃的东方意境，并符合当下的审美形式。有别于一般意义上新东方空间的华丽感和复古性，海联汇更讲究陈设、配置和对商务空间中人文气息的营造，着重于控制空间的品位，在精简传统中式元素的同时，又不失东方意境。

051

　　视野里并不见中式装饰常用的木刻雕花、青花纹理、大红灯笼,但就是那一抹清泉、一张藤椅、一片蒲团让我们感受到其内敛的中式禅意。或许是源于"海联汇"的名称,设计师根据业主的要求,将"水"和"海"的概念作为餐厅设计的主题。在所有象征海、水概念的元素中,设计师以水波的圆弧纹理为灵感,将形态、大小、组合方式不一的"圆"呈现于空间各处。餐厅出口外立面墙上错落镶嵌的各式圆形陶盘装饰,质朴之余,也在传达关于餐饮的信息;大面积的天花和背景墙被刷上圆弧纹理,空调出风口则设计成水纹状的圆形,犹如水波荡漾;入口及大包厢的玻璃表面漆上海水螺旋状,大圆套小圆的效果也被不断重复。公共就餐区的隔断围栏内,白色水平管织成有序的纵向线条,传达雨的概念。此外,以水母、海藻等海底生物为创作原型,进行变体处理的落地灯散落在空间中。蜿蜒的体态和流水般的纹理,整体造型风姿绰约,其藤制工艺在无形中又增添了几分清淡雅致的情趣。

　　新东方的巧妙之处莫过于将传统意境与当代艺术、传统元素与当代手法巧妙融合,设计师恰如其分地表达了这别有韵味的复合式美学。空间的结构通过大体块的拼接构成,用块面搭接的方式穿透延伸。无论是以传统"弓"字形护栏作隔断的公共就餐区,还是用现代玻璃、方钢和木质踏板围合出的透明封闭包厢,抑或是以原木板块打造的隔断屏风、书架和陈列柜,不同体块之间的组合刻意而又自然,构成了极具表现力的功能区域。空间布局讲究每一个细节的搭配,每个单品对于风格的完整都有自身的意义。

　　基于设计师"以国际性的视野,做区域性的文化"的理念,代表老福州文化的装饰被巧妙运用:餐厅酒店内部入口处的背景墙,描绘着昔日大洋百货、中亭街、仓山老城区的素描跃然于上。包间内部,以三坊七巷的旧建筑符号为题材定制的手绘作品占据半壁。设计师对传统文化氛围的渲染也匠心独运:餐厅外部入口以一方青竹、一方泉水、一方陶缸荷叶散发淡淡的闲情雅致;入口内部开辟品茗区,以明式家具的硬朗造型传达质朴、轻松的氛围;由大小不一的原木块组合而成的屏风上刻着唐代古诗《春江花月夜》的节选;原木书架和陈列柜分别放置陶瓷工艺品和传达养生药膳概念的中药材、菌菇标本。而满置葡萄酒的酒文化包间、几张组合拉伸处理的藤制餐椅、室内印象派风格的水墨画作等现代元素则传递出更加多元的审美主张。当意象的东方元素与现代时尚感邂逅,当那些象征各异的古韵气息悄然融入现代工艺感的环境,亦如包容、内敛而低调神秘的个性一般,浓妆淡抹且静水流深。

THE ORCHID PAVILION NO.1
兰亭壹号

The Orchid Pavilion No.1 is located at Binhe East Road, Taiyuan, Shanxi, and it is a high-end dining club which is based on respecting traditional culture and business philosophy of health. Indoor space integrates the elegance and nobility of traditional Chinese culture with modern fashionable creativity as a whole. While using modern design language in interior space planning, designers also apply elements like Chinese classical pane and traditional crafts lacquer cabinets and so on, to refine a form of modern aesthetic taste, making it be the texture of space interface, and lightening a quiet, deep and elegant atmosphere through the light against the background express of space. Water element activates the space. Behind the gurgling trickle water curtain wall is a slated which is carved with Wang Xizhi's The Orchid Pavilion, peaceful but also full of agile sense. At the end of the corridor, the Buddha wall is collocated with metal lotus of modern technology, as a result, the mood is harmonious despite of the material conflicts, echoing a design technique combining tradition with modern techniques.

Design Company 设计公司
Dinghe Architecture and Decoration Design Engineering Co., Ltd. 河南鼎合建筑装饰设计工程有限公司

Designers 设计师
Huafeng Sun, Zhongxun Kong 孙华锋、孔仲迅

Project Location 项目地点
Taiyuan, Shanxi 山西太原

Project Area 项目面积
1,450 m²

Main Materials 主要材料
Black and white jade stone, French wood stone, red oak panel, straw wallpaper, wood grillwork, etc.
黑白玉石材、法国木纹石、红橡面板、草编壁纸、实木花格等

兰亭壹号位于山西太原滨河东路，是以尊崇传统文化及健康养生经营理念为主导的高端餐饮会所。室内空间融中国传统文化的高贵典雅与现代时尚创意于一体。在运用现代设计语汇进行室内空间规划的同时，设计师运用中国古典窗格、传统工艺的漆柜等元素提炼出具现代审美情趣的形式，使其成为空间界面的肌理，透过光线的映衬烘托出静谧、深沉、高雅的空间氛围。而水元素的加入则活跃了空间，潺潺细流的水幕墙背后是雕刻着王羲之《兰亭集序》的石板，静逸中透着灵动。走廊尽头万佛墙与现代工艺制作的金属荷花的搭配，材质冲突中意境却和谐，呼应了传统结合现代的设计手法。

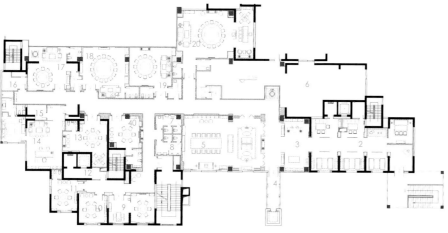

Floor plan
平面布置图

1. Private room
2. Deck area
3. Gift show
4. Elevator
5. Party area
6. Kitchen
7. Female room
8. Male room
9. Private room 9
10. Private room 8
11. Private room 7
12. Storage room
13. Private room 6
14. General manager's office
15. Storage space
16. Storage space
17. Private room 1
18. Private room 2
19. Private room 3
20. Private room 4

1. 包间
2. 卡座区
3. 礼品展示
4. 电梯间
5. 聚会区
6. 厨房
7. 女卫
8. 男卫
9. 包间9
10. 包间8
11. 包间7
12. 储物间
13. 包间6
14. 总经理办公室
15. 储物间
16. 储物间
17. 包间1
18. 包间2
19. 包间3
20. 包间4

会所的陈设设计采用中西合璧的方式，力求通过西式家具的舒适度与中式家具的传统韵味相结合营造私密尊贵且具人文气息的空间气质。墙面悬挂的字画均为大师真迹且经过设计师对比例尺度的严格把控，使其和谐融入空间之中，是整个会所的点睛之笔。精心定制的灯具与空间形成完美的对话，传统绘画复制的漆屏散发着细腻悠远的古韵。

整个空间沉而不闷、透而不散，于无形间流露出浓浓的古典人文气息，表达出对传统文化的敬意与向往。置身兰亭壹号，享受美食与艺术的美妙体验……

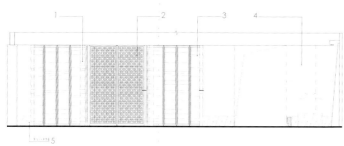

1. Red oak panel closely pieced, stained varnish finishing surface
2. Custom solid wood carving lattice
3. Red oak panel finishing folding door
4. 700x1200mm gray wood-grain stone composite panels closely pieced
5. Black and white jade stone finishes
6. 700x1200mm gray wood-grain stone composite panels closely pieced
7. Custom solid wood carving lattice
8. Black and white jade stone finishes

Elevation of hall
走廊立面图

Elevation of hall
走廊立面图

1. 红橡面板密拼，染色清漆饰面
2. 定制实木雕刻花格
3. 红橡面板饰面折叠门
4. 700mmX1200mm灰木纹石材复合板密拼
5. 黑白玉石材饰面
6. 700mmX1200mm灰木纹石材复合板密拼
7. 定制实木雕刻花格
8. 黑白玉石材饰面

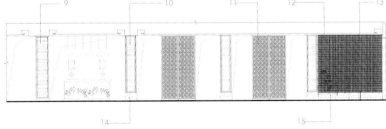

Elevation of deck area
卡座区立面图

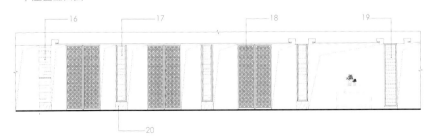

Elevation of deck area
卡座区立面图

9. Custom solid wood wireframe, brocade hard packing
10. Custom solid wood wireframe, oak panel closely pieced stain varnish finishing
11. Custom solid wood carving lattice
12. 80X6mm steel plate black lacquer finishing (spacing 140mm)
13. Sandstone Buddha wall
14. Black-and-white jade stone finishing
15. Black-and-white jade stone finishing
16. Red oak panel closely pieced, stained varnish finishing
17. Custom solid wood frame, oak panel closely pieced, stained varnish finishing
18. Custom solid wood carving lattice
19. Custom solid wood wireframe, brocade hard packing
20. Black-and-white jade stone finishing

9. 定制实木线框，锦缎硬包
10. 定制实木线框，橡木面板密拼染色清漆饰面
11. 定制实木雕刻花格
12. 80mmX6mm钢板黑漆饰面(间距140mm)
13. 砂岩万佛墙
14. 黑白玉石材饰面
15. 黑白玉石材饰面
16. 红橡面板密拼，染色清漆饰面
17. 定制实木框，橡木面板密拼染色清漆饰面
18. 定制实木雕刻花格
19. 定制实木线框，锦缎硬包
20. 黑白玉石材饰面

065

LIU RESIDENCE
刘家大院

Jiangnan residence is a representative of private gardens. After years, a sense of living that has not gone away is brought, imperceptibly relaxing people's body and mind. Because of its natural, comfortable and sunny personality, it becomes a noble example of the traditional architectural layout. The form of the residence inherits the classical living culture.

The prime minister Liu Yong (Liu Luoguo) in Qianlong years has twice served the Jiangsu provincial education commissioner in Jiangyin during Kangxi 59 (1720). Liu residence is the residence of Liu Yong during his tenure in Jiangyin. Liu residence drew materials from the regional culture of Liu Yong's former residence to restore the old house and used the inherent advantages of the construction and environment to create a cultural dining club with modern functions. It inherits and carries on the culture of the former residence and celebrity culture as well as Jiangyin's culture; the original ecological style is combined with fashionable modern design, creating an intermediate region "gray space" which harmoniously consists of the city, architecture, nature and human as a whole.

Designer 设计师
Tianbin Xiang 项天斌
Project Location 项目地点
Wuxi, Jiangsu 江苏无锡
Project Area 项目面积
3,500 m²

　　江南宅院，为私家园林之代表。经历岁月沉淀，带来一份并未远去的居住感，潜移默化间令身心为之舒畅，以其自然舒适，阳光充沛的个性，成为传统建筑形态布局的高尚典范。宅院的形式，传承古典居住文化。

　　乾隆年宰相刘墉（刘罗锅）于康熙五十九年（1720年），曾两次担任江苏学政驻节江阴。刘家大院即刘墉江阴任职期间官邸。刘家大院借助刘墉故居的地域文化，还原其建筑古宅，利用建筑及环境的先天优势，打造具有现代功能的人文餐饮会所。传承故居文化，传承名人文化，传承江阴文化。原生态与时尚的现代设计风格相结合。创造一个集城市、建筑、自然和人和谐共处于一体的中间地带"灰色空间"。

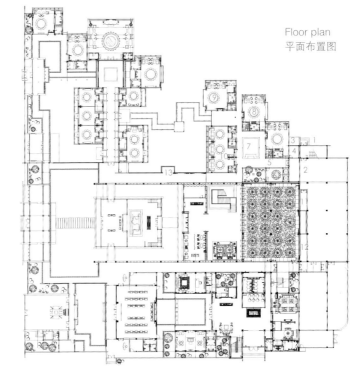

Floor plan
平面布置图

1. Staff entrance
2. Distribution room
3. Flower pool
4. Duty room
5. Gallery
6. Spare room
7. Yard
8. Wing-room 1
9. Wing-room 2
10. Wing-room 3
11. Sound control room
12. Sound control room
13. Viewing deck

1. 员工入口
2. 配送间
3. 花池
4. 值班室
5. 廊
6. 备用厢房
7. 院
8. 厢房1
9. 厢房2
10. 厢房3
11. 音控室
12. 音控室
13. 观景台

Box 1
包间立面图-1

Box 2
包间立面图-2

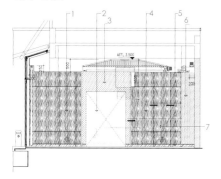
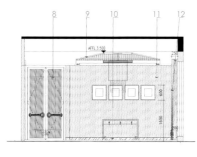

1. Wood finishing door
2. Finished Chinese wardrobe
3. Plain pattern wallpaper
4. Wood finishing
5. 1mm tungsten steel finishing skirting lines
6. Plain pattern wallpaper
7. Wood finishing door
8. Custom bronze door and copper door jacket
9. 30x30mm solid wood lines
10. 1mm tungsten steel finishing skirting lines
11. Plain pattern wallpaper
12. 1mm tungsten steel finishing point
13. Flame-retardant fabric curtain
14. Yarn curtain
15. Matt board jacket of window
16. Plain pattern wallpaper
17. 1mm tungsten steel finishes skirting lines
18. Wood finishing
19. 1mm tungsten steel finishing point
20. Flame-retardant fabric curtain
21. Original glass curtain wall
22. Custom chandelier
23. Yarn curtain
24. LED soft tube lights

1. 木饰面暗门
2. 成品中式衣柜
3. 素纹墙纸
4. 木饰面
5. 1mm钨钢饰面踢脚
6. 素纹墙纸
7. 木饰面暗门
8. 定制铜门及铜门套
9. 30mmx30mm实木线条
10. 1mm钨钢饰面踢脚
11. 素纹墙纸
12. 1mm钨钢饰面收口
13. 阻燃织物窗帘
14. 纱帘
15. 黟县膏亚光板窗套
16. 素纹墙纸
17. 1mm钨钢饰面踢脚
18. 木饰面
19. 1mm钨钢饰面收口
20. 阻燃织物窗帘
21. 原土建玻璃幕墙
22. 定制吊灯
23. 纱帘
24. LED软管灯

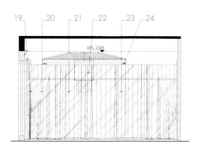

Box 3
包间立面图-3

Box 4
包间立面图-4

YING SHENG BRAND— A FLASH OF COLLISION

盈盛号——碰撞出来的光芒

The traditional old residence with wooden structure is just like a shuttle of time and space in current rational commercial space. When people step into the entrance of Yingsheng Brand showroom, a girl wearing the beautiful costume of the She nationality smiles and shows in front of people, and the wall behind is engraved with "Three Liaos Song", while the ancient silver jewelry which is placed on both sides seems to tell here is one story associated with the She nationality. Moving ahead a few steps, there is aisle like a thin strip of sky. When people lift their sight, the galaxy appears. The starry sky and the ancient wall together with a line of wild geese flying to the moon make people can't help but thinking of the scene when the wild geese come back, the moon is shining the west floor.

Design Company	设计公司
Fuzhou Zhonghe Design	福州中和设计事务所
Designer	设计师
Ruifeng Chen	陈锐峰
Project Location	项目地点
Fuzhou, Fujian	福建福州
Project Area	项目面积
300 m²	
Main Materials	主要材料
Blue stone, tiles, black dragon marble, optical fiber, etc.	青石板、瓦片、黑金龙大理石、光纤等

传统木结构老宅在当下理性的商业空间里犹如时空梦幻般的穿梭。当步入盈盛号展示店入口一刹那,一个穿着畲族精美服饰的姑娘微笑展现在人们眼前,背后的那面墙刻着"三寮曲",两边摆放的古银饰品,仿佛在诉说这里是一个与畲族相关的故事。前行几步,有似一线天的过道,抬眼望,银河乍现,点点繁星,古朴墙面,一行大雁飞向圆月那边,不禁让人联想"雁字回时,月满西楼"的情境。

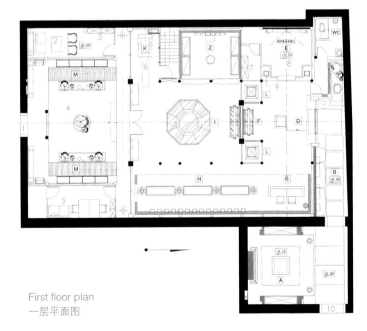

First floor plan
一层平面图

1. Backdoor of tower lane
2. Reception room
3. Chinese style book shelves
4. Open-air landscape
5. Office
6. Wagon model
7. Dressing room
8. Performance area
9. Decorative stone animals
10. Main entrance of south backstreet

1. 塔巷后门
2. 接待间
3. 中式书架
4. 露天景观
5. 办公室
6. 马车模型（满载而归）
7. 更衣间
8. 表演区
9. 装饰石兽
10. 南后街主入口

前方有流水自高处向下流至水池,似乎银河倾泻,一道感应门,已到展厅内,映入眼前的是空间上古典与时尚相映生辉,高挑的地方特色古民宅,精美绝伦的银饰品,在灯光下银光闪闪,琳琅满目,空中星光灿烂,光影迷离。中心区内有乾坤,先天八卦展示台,此元素来自于畲族经典银饰品之一的八卦"长命富贵锁",不仅满足了功能要求,又提炼了元素在空间中的灵气,上方以古银为材料的女娲手托五彩石,腾空飞起,把人的思想一下延伸到了华夏文明的启蒙时期,游离在此空间里,不经意间就产生了想拥有一件盈盛号民族银饰品的冲动。

HUA JIAN TANG——THE SEASON YARD IN ZHOU ZHUANG

花间堂·周庄季香院

According to the client's demand, this boutique hotel will be combined with Zhouzhuang's picturesque scenery and history to reflect the ancient village's unchanged quiet and elegant life and also retain and continue local history and culture.

In order to correspond to Zhouzhuang's sense of history and Chinese cultural elements, the designer Thomas DARIEL who loves Chinese culture sets the design theme of this boutique hotel to be "sensory journey going through the seasons", whose inspiration comes from the traditional Chinese 24 seasonal segments. In China, according to the farming activity habits in ancient times and the laws of nature, people divided one year into twenty-four segments, such as the Beginning of Spring, Grain Rain, Light Snow and so on according to the position of the sun in the ecliptic.

Design Company 设计公司	
Dariel Studio	
Designer 设计师	
Thomas Dariel	
Project Location 项目地点	
Zhouzhuang, Jiangsu 江苏周庄	
Project Area 项目面积	
2,500 m²	

　　根据客户的要求，此精品酒店将与周庄如画风景和历史相结合，体现这古镇古往今来一直未变的恬静而优雅的生活，保留并延续当地的历史文化。

　　为了契合周庄这一古镇的历史感及中国文化元素，热爱中国文化的设计师Thomas Dariel 将这个精品酒店的设计主题设定为"穿越季节的感官之旅"，其灵感来自于中国传统的二十四节气。在中国，依照古时农耕的作息习惯和自然规律，人们根据太阳在黄道上的位置把一年平分为二十四个节气，如立春、谷雨、小雪等。

　　花间堂·周庄季香院酒店位于江南水乡——周庄,粉墙黛瓦,厅堂陪弄,临河的蠡窗,入水的台阶,在这里,千年的历史也隐在江南迷迷茫茫的烟雨中,其温婉绰约的神韵随着碧波在不经意间一波一波地荡漾开来。离上海仅1.5小时的车程使其成为上海背后避世休闲的绝好去处。

　　这个精品酒店项目是由三幢明清风格的老建筑改造而成,相传这三幢老建筑曾分别属于戴氏一家兄弟三人,各自在这宁静的小镇经营其营生。时至今日,经过岁月的年轮无情地碾过,在改造之前这三幢独栋建筑分别被用做博物馆、茶室、客栈,并有一部分已经废弃。Dariel Studio非常小心地对这些优秀的古建筑进行修复并将其合并改建成拥有20套客房的精品酒店,同时希望可以保留建筑最原始的空间结构以及其历史传承。

　　昔日的戴宅被分为东、西、中三宅,三宅独立而建,却又紧贴相连,成为一个整体,格局迥异,各具特色。Dariel Studio用了近半年的时间对其进行修复改建,包括地面高低的统一,主梁的加固,门窗的修复和重建,结构的重新划分等。

　　首先,在酒店各房间的布局上,根据日照的上升降落的分布规律,自南向北将春夏秋冬依次在各排房间进行演绎。从浅浅的大地色,到跳跃的橘色,过渡到深沉的紫色,演绎了四季的不同个性。采用四季迥异的花卉来命名不同的客房名,芷樱、碧荷、丹桂、墨兰……并且运用不同的软装和灯饰来诠释,仿佛终于找到了隐世的桃花源,置身于四季的国度,久久不愿离去。

　　其次,设计师选取了几个重要的节气进行分别表现,使整个酒店的空间立体分布具有季节性的标志:春分,春暖花开,岸柳青青。用这一个节气来表达带领来访者进入一种新鲜的入住体验的接待处,那是最恰当不过了。春分,昼夜平分之意,也能恰如其分地代表接待处这一贯通里外的作用。芒种,麦子丰收,酒醇得正是时候的季节,也是如同红酒一般令人珍惜的,将红酒吧用芒种来形容是再贴切不过了。夏至和冬至,色彩的强烈对比和西式吧台与中式家具的搭配,表现出中西餐厅美食文化的激情碰撞。惊蛰,用于表现适合冥想的地方——阅读室,在这里聆听自己的内心,幡然醒悟更多哲理。悬于走廊的笼状灯笼、淡金的配色、舒适的沙发,无一不使你希望在这个电闪雷鸣、虫儿苏醒的节气窝在这里品着香茗,读本好书。白露和小暑,分别代表水吧和茶室。白露象征了水的洁净与润泽;小暑正显示了茶所需要的温度。在这里,你可以体验不同季节带来的非凡感官体验,感受令人颤栗的时光之旅。

为了更好地保护当地文化和传统建筑，在进行修复改建时，小到一砖一瓦一石子都被编号保留起来，并且修旧如旧，重现新生。那些实在被毁坏得严重的，Dariel Studio采用相同形状花纹进行重新制作以符合明朝风格。在庭院门楼上秀才陶惟坻的题字"花萼联辉"，寓意戴氏兄弟连心，必能兴旺家业；堂楼之间天井相连，雕梁画栋，颇有气魄。正厅对面为三进重檐风火墙门楼，题额刻有"剡溪遗泽"，意为怀念故乡情深；房间里也随处可见留下的弯曲的悬梁和雕花题刻。除了保留原有的能够被保留下的，更多的仿古装饰被装点进了这个空间以切合整体的环境。灵感来源于中式手提食盒的各类柜子，竹子形状为支柱的中式大床，古式的门把，卫浴间门上的雕花铜片，镶在墙上的民族项链，各式的中国瓷器花瓶，亮色的交椅，墙上用各色毛笔笔刷高低错落的排列作为装饰……

除了保留和恢复其中式的特点，来自法国喜爱将中法文化结合在一起的Thomas Dariel也将中西融合在这个室内设计中表现得淋漓尽致。代表惊蛰节气的阅读室里，配有一个西式壁炉和小型钢琴，不但在中式的氛围中更添一份温暖，更让人有种置身于法国文艺复兴时期的感觉，如此取长补短的结合真是恰到好处。中西餐厅的强烈色彩对比，巨大的悬式吊灯，以及那引人注目的法式花饰瓷砖砌成的吧台让人在一片传统中找到新鲜亮眼之处。更特别的是，装点墙面的各种画像也充分向中国丰富的手工艺品致敬。同时，摄影作品《水中的墨滴》表现出一种结合中国传统水墨书法以及当代的诗歌的感觉。各种中式装饰以及西式装饰的互相混搭营造出现代与古典的别样完美的结合。

在不同的公共空间，皆会产生不同的体验之旅。作为一个精品酒店，当然不同于传统的五星级酒店，在这里，你可以全身心地放松并且如同在家般自在地度假。这里有红酒吧、阅读室、影音观摩室、水吧、水疗、瑜伽室等活动中心满足休闲享受的需求。每个空间都将带您通往享受之旅。充分利用空间与光线、展现本土文化与传统、精心选材并钟情于艺术与手工艺品，Dariel Studio通过种种手段使朴实与典雅交相辉映。这一空间的主要设计概念是永恒之时、静谧之美与精妙奢华。我们笃信舒适并非在于区区的富贵，而是应当体现在精致与珍奇之中。花间堂·周庄季香院酒店的室内设计融合了传统与现代。在这幢典型的19世纪建筑中，我们将古老中国的特色与现代摩登家具交织在了一起，伴着当地特色的手工艺品，使游客沉浸于这座城市的灵魂与个性之中，并在现代的视觉效果与精致优雅的氛围下恢复活力。

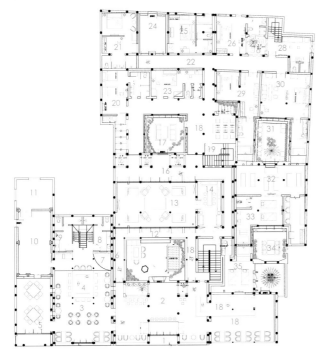

1. Hotel entrance	13. Reading and listening room	25. Guest room
2. Art space	14. Wine bar	26. Guest room
3. Dining room	15. Arch	27. Patio
4. Bar area	16. Corridor	28. Guest room
5. Dining room	17. Patio	29. Guest room
6. Corridor	18. Internet surfing	30. Guest room
7. Stage	19. Storage room	31. Patio
8. Ladies' room	20. Guest room	32. Yoga room
9. Men's room	21. Guest room	33. SPA room 1
10. Kitchen	22. Corridor	34. Patio
11. Strong motor room	23. Guest room	35. SPA room 2
12. Corridor	24. Patio	36. Patio

1. 酒店入口	13. 阅读、静听室	25. 客房
2. 艺术空间	14. 红酒吧	26. 客房
3. 餐厅	15. 拱门	27. 天井
4. 吧台区	16. 走廊	28. 客房
5. 餐厅	17. 天井	29. 客房
6. 走廊	18. 网际冲浪	30. 客房
7. 舞台	19. 储藏室	31. 天井
8. 女厕所	20. 客房	32. 瑜伽室
9. 男厕所	21. 客房	33. SPA房间1
10. 厨房	22. 走廊	34. 天井
11. 强电机房	23. 客房	35. SPA房间2
12. 走廊	24. 天井	36. 天井

First floor layout
一层平面图

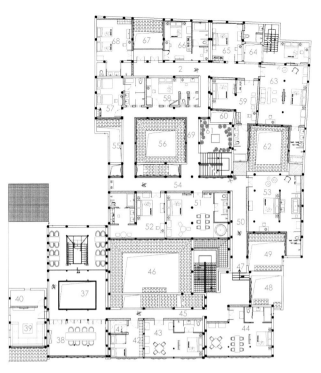

Second floor layout
二层平面图

37. Shared space	54. Aisle
38. Meeting room	55. Patio
39. Audio-visual room	56. Patio
40. Strong motor room	57. Guest room
41. Storage room	58. Guest room
42. Staff room and straw room	59. Guest room
43. Guest room	60. Patio
44. Guest room	61. Sky garden
45. Aisle	62. Patio
46. Patio	63. Guest room
47. Arch	64. Patio
48. Patio	65. Guest room
49. Patio	66. Guest room
50. Aisle	67. Patio
51. Guest room	68. Guest room
52. Guest room	69. Aisle
53. Guest room	

37. 共享空间	54. 过道
38. 会议室	55. 天井
39. 影音室	56. 天井
40. 强电机房	57. 客房
41. 储藏室	58. 客房
42. 员工室和布草间	59. 客房
43. 客房	60. 天井
44. 客房	61. 空中花园
45. 过道	62. 天井
46. 天井	63. 客房
47. 拱门	64. 天井
48. 天井	65. 客房
49. 天井	66. 客房
50. 过道	67. 天井
51. 客房	68. 客房
52. 客房	69. 过道
53. 客房	

IMPRESSION OF FEET-MASSAGE
印象足道

Designers make the intention of water as the theme of space, echoing to the property in shape and spirit and cleverly symbolising the intention of water. The yellow and white natural color step by step renders elegant and noble, comfortable and cozy space feeling. Through the cutting techniques of spatial structure, designers make the rhythmic movement of line styles in harmony with all the functional spaces. The balance of the upper, middle and lower parts, the contrast between the virtual and real space, the match of light, simple and unadorned parts, the interaction of entity and light and shadow create a deep Zen flavor of Southeast Asia. In interpretation of elements, the wave with relief sculpture sense on wall is not only a contemporary performance in form, but also distinctly suggests the theme of intention of "water". The reed curtain, ink color silk yarn and misty classical landscape curtain not only add cultural temperament to space, but also create a visually and psychologically rich feeling in the materials and form, which creates a smooth and elegant, leisurely and warm space.

Design Company 设计公司
S-zona Designer Consultant Inc.　无锡市上瑞元筑设计制作有限公司
Designer 设计师
Liming Sun　孙黎明
Design Team 参与设计
Shunfeng Geng, Hao Chen, Hengsong Zhu　耿顺峰、陈浩、朱恒松
Project Location 项目地点
Yangzhou, Jiangsu　江苏扬州
Project Area 项目面积
1,000 m²
Main Materials 主要材料
Italian wood stone, the magnificent stone acid washing stone, straw wallpaper, northeast China ash opening paint, ect.
意大利木纹石材、金碧辉煌石材酸洗面石材、草编墙纸、水曲柳开口漆等

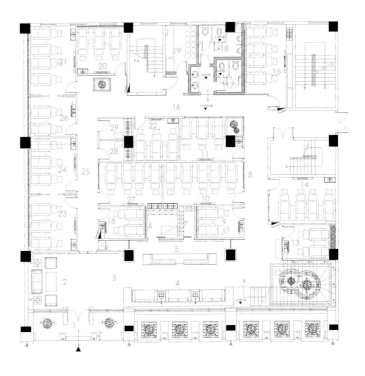

First floor layout
一层平面布置图

1. Main entrance
2. Waiting area
3. A lobby
4. A area to change shoes
5. Reception desk
6. Office
7. Deposit shoe room
8. L-shaped corridor
9. Box for two persons
10. Box for five persons
11. Box for three persons
12. Landscape platform
13. Box for two persons
14. Box for four persons
15. Box for four persons
16. K corridor
17. Woman toilet
18. Man toilet
19. Towel room
20. Box for three persons
21. Box for four persons
22. Box for three persons
23. Box for three persons
24. Box for three persons
25. Lcorridor
26. Box for two persons
27. Box for four persons
28. Woman dressing room
29. Man dressing room

1. 主入口
2. 等候区
3. A大厅
4. 换鞋区
5. 服务台
6. 办公室
7. 存鞋室
8. J走道
9. 2人包厢
10. 5人包厢
11. 3人包厢
12. 景观平台
13. 2人包厢
14. 4人包厢
15. 4人包厢
16. K走道
17. 女卫
18. 男卫
19. 毛巾间
20. 3人包厢
21. 4人包厢
22. 3人包厢
23. 3人包厢
24. 3人包厢
25. L走道
26. 2人包厢
27. 4人包厢
28. 女宾更衣
29. 男宾更衣

　　以水的意向作为空间主题，与液态的属性保持形神的呼应，并对水的意向做了巧妙的符号化处理，赫黄白的自然化色彩渐进渲染出清雅高贵、舒适惬意的场所感受。设计师通过空间结构的切割手法，赋予了线形的律动与各功能空间的和谐衔接，上中下的均衡、虚实空间的对比、轻巧与朴拙的搭配、实体与光影的互动，营造出禅味深长的东南亚意蕴。在元素的演绎上，墙体浮雕感的水波即为形式上的当代化表现，又鲜明暗示着"水"这一主题意向；苇帘、水墨的绢纱、飘渺的古典山水挂帘，不仅平添了空间的文化气质，又在材质与形式表现上创造了视觉与心理上的丰富感受，造就了一个净爽淡雅的休闲温馨的空间。

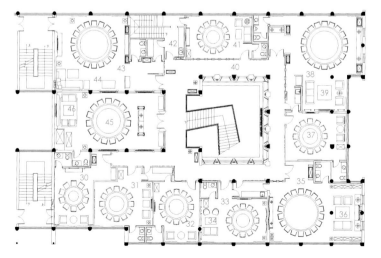

Second floor layout
二层平面布置图

30. Box for ten persons	30. 10人包厢
31. Box for twelve persons	31. 12人包厢
32. Box for twelve persons	32. 12人包厢
33. Box for twelve persons	33. 12人包厢
34. Resting area	34. 休息区
35. Luxurious box for eighteen persons	35.18人豪华包厢
36. Resting area	36. 休息区
37. Box for twelve persons	37. 12人包厢
38. Luxurious box for eighteen persons	38. 18人豪华包厢
39. Resting area	39. 休息区
40. 2L corridor	40. 2L走道
41. Box for twelve persons	41. 12人包厢
42. 2M public pantry	42. 2M公共备餐间
43. 2A Luxurious box	43. 2A豪华包厢
44. Resting area	44. 休息区
45. Box for fifteen persons	45.15人包厢
46. Resting area	46. 休息区

LUCKY AND PEACEFUL FOR A LIFETIME
祥和百年

The designers adopt the simple Chinese technique to perfectly combine implication with casual natural temperament in this case. Chinese concise style is used in design alternately, not showing an unexpected sense, but presenting a kind of compatible beauty. Through the change of geometrization, visualization, contrast and rhythmization, the so-called classical vocabulary is transformed. Designers simplify the lines in line, particular pay attention to a sedate and extravagant single color in color and advocate massiness, solemnity, texture, luxury, and even "straight to the point" in spatial language spread.

Design Company 设计公司
Hefei Xu Jianguo Architectural Interior Decoration Design Co., Ltd. 合肥许建国建筑室内装饰设计有限公司

Designer 设计师
Jianguo Xu 许建国

Design Team 参与设计
Tao Chen, Kun Ouyang, Yingya Cheng 陈涛、欧阳坤、程迎亚

Project Location 项目地点
Hefei, Anhui 安徽合肥

Project Area 项目面积
1,600 m²

Photographer 摄影师
Hui Wu 吴辉

Main Materials 主要材料
Northeast China ash wood finish, barry grey granite, grey bricks, antique floor tiles, ect.
水曲柳木饰面、芝麻灰花岗岩、小青砖、仿古地砖等

　　外立面采用中式园林的手法给人眼前一亮的感觉，显得很整体大气，入口则用了照壁的方式，将曲径通幽、移步换景的意境表达了出来，室内也控制得相当得当，整体的节奏以及材料和色彩都把握得很到位，女子十二乐坊的元素运用得很生动。

　　设计师立足中国本土文化，把徽文化与中式文化完美结合，运用了多种表现形式。设计的某些部分是徽文化的剪辑，设计师借用《兰亭序》，表现了本案的设计意境，设计界现有众多设计风格，如欧式，东欧，泰式等。在本案中设计师着力打破设计现状，找到属于中国人自己创新的设计风格，真正体现中国人自己设计中的简洁、简约、儒雅的精髓，打造具有诗人情怀的空间。在部分包厢设计当中，就餐区与休息区合理分开，休息区设有很高的天井，此环境适合大家畅所欲言，饭前吟诗作乐，欣赏天空的月亮，大有"举头望明月，低头思故乡"之情。

　　"这一天，天气晴朗，和风轻轻吹来。向上看，天空广大无边，向下看，地上事物如此繁多，这样来纵展眼力，开阔胸怀，穷尽视和听的享受，实在快乐啊！"这句话是本案想体现的完美意境吧。

本案以简洁的中式手法，将含蓄内敛与随意自然两种气质完美结合。中式简约风格在设计中穿插运用，不显突兀，反而呈现一种兼容并蓄的美。通过把所谓古典语汇几何化、图像化、对比化、节奏化等方式转化，在线条上化"繁"为"简"；在色调上，讲究稳重而贵气的单一色彩；在空间语言传播上，主张厚重、庄严、质感、奢华，甚至是"开门见山"。

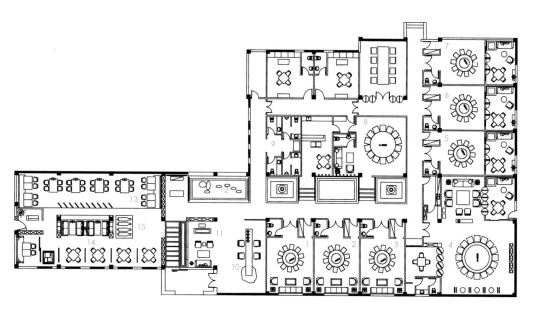

Floor plan
平面布置图

1. Box — 1. 包厢
2. Box — 2. 包厢
3. Box — 3. 包厢
4. Luxurious box — 4. 豪华包厢
5. Box — 5. 包厢
6. Box — 6. 包厢
7. Box — 7. 包厢
8. Luxurious box — 8. 豪华包厢
9. Toilet — 9. 卫生间
10. Reception desk — 10. 前台
11. Reception area — 11. 接待区
12. Landscape — 12. 景观
13. Extra seat — 13. 散座
14. Extra seat — 14. 散座
15. Landscape — 15. 景观

IMPRESSION OF HAKKA
印象客家

Any kind of culture or idea should be cultivated through a carrier, developing and flourishing subsequently. "Impression of Hakka" is just such a place. It is full of imagination under the designer's careful planning, and creates a lot of meaningful situations between the visible and invisible. Therefore, when we have dinner or drink tea here, what we experience is not only a senior enjoyment of the taste bud, but also a process of touching the soul. "Impression of Hakka" is located in the A-ONE Sports Park, and the location hidden in the depth adds a bit low-key and restrained sense to the dining space. The cultural idea "Tracing the source, making one's home everywhere" is also imperceptibly interpreted.

Design Company 设计公司
Fuzhou Damu Heshi Design　福州大木和石设计联合会馆
Designer 设计师
Chen Jie　陈杰
Project Location 项目地点
Fuzhou, Fujian　福建福州
Project Area 项目面积
1,100 m²
Photographer 摄影师
Yuedong Zhou　周跃东
Text Editing 文字
Yongxiao Jiang　江雍箫
Main Materials 主要材料
Old wood, linen, acrylic panels, rice paper, gray mirror, clouds lights, ect.
旧木、麻布、亚克力板、宣纸、灰镜、浮云灯等

　　任何一种文化、一种理念，都要通过一个载体来培养，继而发扬光大。"印象客家"便是这样一个地方，它在设计师的精心规划之下充满了想象，于有形无形之间塑造出许多耐人寻味的情境。于是，我们在此用餐或品茶，体验到的不仅是味蕾的高级享受，更是触动心灵的一个过程。"印象客家"位于A-ONE运动公园内，隐于深处的位置给这个餐饮空间增加了几分低调与内敛。"追根溯源，四海为家"的文化理念也在潜移默化中得到些许诠释。

"印象客家"的门面上方用斑驳的铁皮做装饰，粗犷的纹理显得厚实而有力量感。下方的圆窗位置，摆放着石磨与擂茶饼，墙面上的地图指示出客家族群在国内的分布情况，这些与客家文化一脉相承的物件在这古朴的空间中悠悠不尽。

尚未进入空间内部，外面的庭院景观已然吸引了我们的目光。曲径有秩的布局丰富了视觉的层次，得益于此，设计师在这个环境中设置了若干包厢。包厢置于自然的怀抱之中，食客便拥有了广阔的视野。同时，玻璃墙面使得窗外郁郁葱葱的景致成为一道天然的背景。渐渐地，这里的一草一木、一砖一瓦，不管是有生命的还是没生命的，都找到了与空间沟通共融的方式。

当我们把视线移到印象客家的主体内部空间时，一种既陌生又熟悉、既单纯又丰富的视觉感受油然而生。用铁锈色的瓷砖铺陈出的空间地面，孕育着不可或缺的气度，并加强了走道的纵深感。中式家具、器皿以及大体量的木柱，适时地分布在相应的位置，让人们在繁简交错之间找到最舒适的体验。品茶与用餐是这里的两个功能区域，虽然它们鲜明地分布在空间的不同位置，但格调的融合让二者的过渡自然而然。沉稳的色调营造出一个幽静淡雅的环境，烙有年代印记的摆设品错落分布其中，没有了浮躁，不见了仓皇。包厢的设计遵循开放式的结构，通过一些巧妙的设计与其他区域进行有效的连接与视觉沟通。在细节的方面，笔触不多的勾勒，却很到位。黑色簸箕印上百家姓，重重叠叠地分布在过道等区域的上方。既是装饰，也蕴含着客家的迁徙文化。此外，一面展示客家建筑的黑白照片墙则把人们带回那悠远的记忆。

通往二楼的楼梯旁是一个矩形的水景，陶缸、睡莲以及周边做旧的墙面，它们搭配出一派宁静诗意。这个区域通过一盏造型别致的藤灯向上贯穿至二楼，上下两个空间便有了交流的可能。在光与影的交互作用下，这个空间的体量变得清晰并获得了形式界定。此外，光还成为一种媒介，让每个物件与其周围的环境产生了对话。水景旁的走道可以到达楼梯的位置，青石踏板的设计增强了空间的延伸性。在靠近水景的墙面上，方形的镂空设计像是一个个取景框，使得空间更具有张力。

进入二楼的区域，玄关墙上的石壁彰显着客家的文化特性。石壁两侧分别是饮茶区与就餐区，

中间的隔断便是从一楼水景处延伸上来的部分。两个功能区域之间可以相互借景，而不是硬性的区分。就餐的包厢中，中式的餐椅与西式的吊灯并不冲突，反而有种浪漫的意味。墙面上的一幅廊桥图案的喷绘作品，又再次向宾客展示了客家文化的建筑形态。与此同时，画面中近大远小的透视效果，恰好为空间增强了纵深感。

印象客家虽是一个质朴的空间，但这种质朴并非奢华的对立面，而是一种平凡的表象，骨子里却充满了丰富的情愫。当阳光顺着周遭的树木摇曳而下，而后暖暖地映在包厢中，不用过多地修饰，如同一张讲述闲适情怀的电影海报。而夜幕降临时，在暖色的灯光映射下，空间便富有了诗意，充盈着迷人的气息。

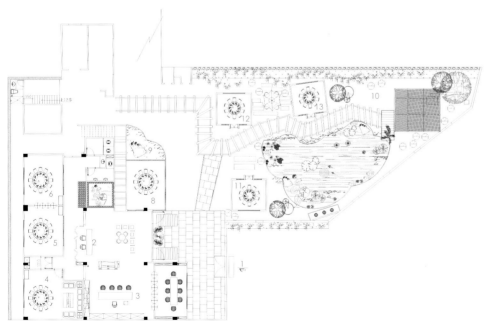

Floor plan
平面布置图

1. Entry — 1. 入口
2. Bar — 2. 吧台
3. Tea zone — 3. 品茶区
4. Box — 4. 包间
5. Box — 5. 包间
6. Box — 6. 包间
7. Landscape — 7. 景观
8. Box — 8. 包间
9. Landscape — 9. 景观
10. Outdoor landscape — 10. 户外景观
11. Box — 11. 包间
12. Box — 12. 包间
13. Box — 13. 包间

NATIONAL YARD
国民院子

The base of this case is located in the reception center of Dalian, China. The case is positioned as the new National nobility to create its particularity and value. Through the clip and blend of Eastern and Western culture, the high and low level and the crossing of inside and outside of the real lines, and the parabola's attachment to the building, the virtual and actual of the building, the designers interpret the characteristics of the new nationals nobility, and the splendid dialogue between space and urban context is derived from them.

Designers constitute the decoration and the spirit lines of space with the local native material, making the environment fully stretch while deeply breathe. Designers show the most clumsy but delicate Montage spirit through natural calligraphy lines.

Totem and materials are voluntarily combined through the lighting, making the wall become an exquisite art which is full of historical charm. The presentation of the Western construction sequence reflects the Montage technique which is a combination of the eastern and western culture.

Design Company 设计公司
Tien Fun Interior Planning Co., Ltd. 天坊室内计划有限公司
Designer 设计师
Qingping Zhang 张清平
Project Location 项目地点
Dalian, Liaoning 辽宁大连
Project Area 项目面积
720 m²
Photographing 摄影
Junjie Liu 刘俊杰
Main Materials 主要材料
Guanyin stone, tawny glass, wallpaper, paint, etc. 观音石、茶镜、壁纸、漆等

本案是位于中国大连的接待中心专案，为创造其特殊性与价值性，设计师以新国民贵族为其定位。设计者透过东西文化的剪辑与交融，实质线条的高低、内外交错，以抛物线依附量体的概念，建筑的虚与实，诠释新国民贵族特色，并衍生出空间与城市脉络的精彩对话。

以当地的原生素材，构成装饰与空间精神线条，让环境氛围尽情伸展同时深度呼吸，以自然的书法笔触线条，呈现大巧若拙的蒙太奇精神。

将图腾与材质透过灯光自行结合，让柱面成为充满历史韵味的精美艺术，西式建筑序列的呈现，体现东西融合兼容并蓄的蒙太奇手法。

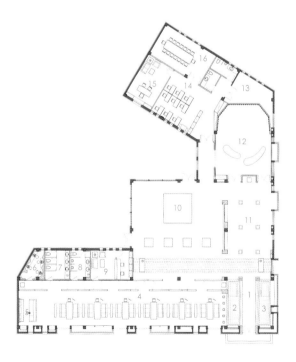

Floor plan
平面布置图

1. Reception area   1. 接待区
2. Waterscape   2. 水景
3. Waterscape   3. 水景
4. Negotiation area   4. 洽谈区
5. Bar   5. 吧台
6. Children's playroom   6. 儿童游戏室
7. Woman toilet   7. 女洗手间
8. Man toilet   8. 男洗手间
9. Financial Room   9. 财务室
10. Big model sets   10. 大模型台
11. Recreational area   11. 休闲区
12. Multi-function room   12. 多功能室
13. Storage room   13. 储藏室
14. Office   14. 办公室
15. Manager's Office   15. 经理室
16. Meeting rooms   16. 会议室

　　大厅与展演厅天花板的处理，设计者将中国人祁愿幸福、仰望光明、迎接希望与温暖的心愿，文人贵族内心逍遥自在的奔放，透过新式素材、序列的拼接手法，将蒙太奇式智慧引渡到空间之中。

　　展台以自然肌理的真材实料，呈现让人心安淡定的空间质感，解开东西方共同的简约设计符码，喜极净，净水无波，明心见性。让参观者能品赏蒙太奇式的豁达与宏观。

SUZHOU GARDEN NO.1
苏园壹号

The original building was a modern-style sales center with a two-storey structure. The designers deliberately design the entrance to let the guests go through the courtyard into the corridor and then into the clubhouse, making guests feel the implication and delicacy of the Chinese garden. The reconstructed building is a kind of simple Huizhou style architectural form. In order to interpret the hall version of "The Peony Pavilion" in the clubhouse, designers use the tenon wood structure of Chinese traditional Huizhou style architecture in the atrium which is also used as seat area, and the floor can be lifted to be used as a stage in case of performances and activities. Private dining rooms' design relies on bookshelves to reflect the feelings of the Chinese literati, and passes the elegant temperament of the clubhouse with great beauty, elegance, and implication. In the furnishings design, the mix and match of Western neo-classical and new Chinese furniture is used, which both brings good comfort and achieves excellent visual effects. The use of navy blue linen perfectly balances the warm tone of a large area of wood color rattan weaves, and curtain and bamboo shade make guests both get a sense of enclosure and break the stiff feature of traditional wooden architecture in the bulk area.

Design Company　设计公司
Dinghe Architecture and Decoration Design Engineering Co., Ltd.　河南鼎合建筑装饰设计工程有限公司
Designers　设计师
Shiyao Liu, Huafeng Sun　刘世尧、孙华锋
Project Location　项目地点
Zhengzhou, Henan　河南郑州
Project Area　项目面积
1,510 m²
Photographing　摄影
Huafeng Sun　孙华锋
Main Materials　主要材料
White wood grain stone, the ancient wood grain stone, African sandalwood matting, hard pack, citywall bricks, etc.　白木纹石材、古木纹石材、非洲紫檀、席编、硬包、城砖等

　　原建筑为两层结构现代风格的售楼中心，设计师将会所入口设计为通过庭院进入廊道再进入会所，让客人有节奏地感受到中国园林的含蓄和精致。改造成的建筑为简约的徽派建筑形式。为了将厅堂版《牡丹亭》在会所中演绎，设计师在兼做散台区的中庭采用了中国传统徽派建筑的榫卯木结构，并将地板做成升降地板以备演出、活动之用。包间的设计以书架为依托体现出中国文人雅士的情怀，以大美、素雅、含蓄来传递出会所的风雅气质。在陈设设计中，设计师采用西方新古典及新中式家具的混搭，既有了很好的舒适感，又取得了极佳的视觉效果。宝蓝色布草的运用使大面积木色藤编的暖色得到很好的平衡，布幔和竹帘让在散台区的客人既有了围合感又打破了传统木构建筑的生硬感。

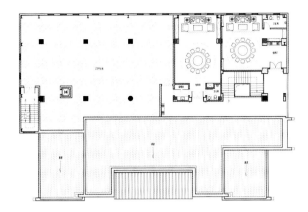

The general layout of the third floor
三层总平面图

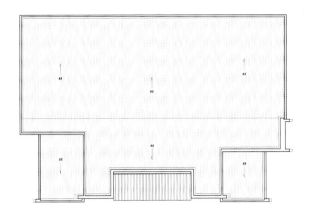

The general layout of the forth floor
四层总平面图

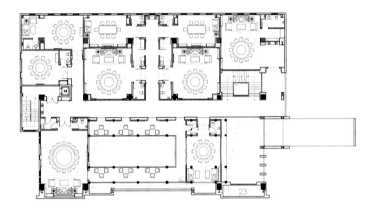

The general layout of the second floor
二层总平面图

1. Outdoor ground	13. Box 1
2. Ramp	14. flower beds
3. Flower beds	15. Box 2
4. Waterscape	16. Woman toilet
5. Waterscape	17. Man toilet
6. Flower beds	18. Box 3
7. Entrance	19. Box 4
8. Lobby	20. Box 5
9. Reception desk	21. Box 6
10. Storage room	22. Box 7
11. Custom stainless steel vertical ladder	23. Storage spaces
12. Auditorium	

1. 室外地面	13. 包间1
2. 坡道	14. 花池
3. 花池	15. 包间2
4. 水景	16. 女卫
5. 水景	17. 男卫
6. 花池	18. 包间3
7. 入口	19. 包间4
8. 大堂	20. 包间5
9. 服务台	21. 包间6
10. 储物间	22. 包间7
11. 定制不锈钢直梯	23. 储物间
12. 演艺厅	

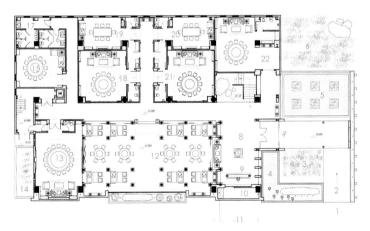

The general layout of the first floor
一层总平面图

"良辰美景奈何天，便赏心乐事谁家院？"几年前欣赏《牡丹亭》的那晚，设计师仿佛穿越了时空，步入了一个江南的庭院，目睹了一段六百年前的凄婉缠绵的爱情故事……几年后再做苏园，这大美的印象便成了此次设计的主题。把声音与建筑，餐饮与文化很好地融合，承载出另一种文化餐饮消费方式，使它成为一种文化景观，一种城市记忆……

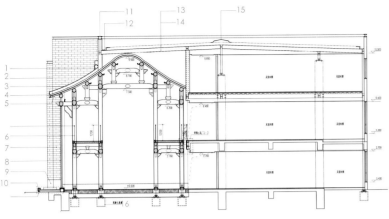
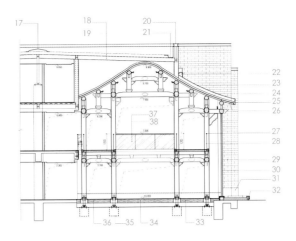

Section plan
剖面图

1. Back
2. Plain tile
3. Waterproof layer
4. Concave tile
5. Eaves
6. Custom solid wood handrails
7. 8+a+8 tempered glass
8. 300X800mm citywall bricks
9. 8+a+8 tempered glass
10. Quartzite stone finishes

11. Pools
12. Cornice made of steel plates
13. Drains made of steel plates
14. Corrugated insulation board
15. H section
16. Water treatment by using steel plates
17. Water treatment by using steel plates
18. Corrugated insulation board
19. H section
20. Cornice made of steel plates

1. 脊背
2. 板瓦
3. 防水层
4. 底瓦
5. 檐子
6. 定制实木扶手
7. 8+a+8钢化玻璃
8. 300mmx800mm城砖（磨砖对缝处理）
9. 8+a+8钢化玻璃
10. 青石板石材饰面

11. 水池
12. 钢板做檐口
13. 钢板做排水沟
14. 瓦楞保温板
15. H型钢
16. 钢板做防水处理
17. 钢板做防水处理
18. 瓦楞保温板
19. H型钢
20. 钢板做檐口

21. Drains made of steel plates
22. Back
23. Plain tile
24. Waterproof layer
25. Concave tile
26. Eaves
27. Custom solid wood handrails
28. 8+a+8 tempered glass
29. 300X800mm citywall bricks

30. 8+a+8 tempered glass
31. Quartzite stone finishes
32. Pools
33. Lifting platform
34. Lifting platform
35. Concrete foundation
36. Lifting platform
37. Custom solid wood handrails
38. 8+a+8 tempered glass

21. 钢板做排水沟
22. 脊背
23. 板瓦
24. 防水层
25. 底瓦
26. 檐子
27. 定制实木扶手
28. 8+a+8钢化玻璃
29. 300mmX800mm城砖（磨砖对缝处理）

30. 8+a+8钢化玻璃
31. 青石板石材饰面
32. 水池
33. 升降平台
34. 升降平台
35. 混凝土基础
36. 升降平台
37. 定制实木扶手
38. 8+a+8钢化玻璃

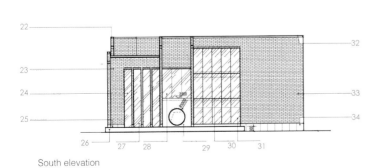

West elevation
西立面图

South elevation
南立面图

1. Cornice made of steel plates
2. 300X800mm citywall bricks
3. New heat-saving aluminum windows
4. New red copper security door
5. Quartzite stone finishes
6. Flower beds
7. Cornice made of steel plates
8. Plain tile
9. Back
10. Cornice made of steel plates
11. 300x800mm citywall bricks
12. Dry-hang dark aluminum veneer
13. Copper plate
14. Flower beds
15. Ramp
16. Black gold stone finishes
17. Quartzite stone veneer
18. Quartzite stone finishes
19. Pools
20. 8+a+8 tempered glass
21. Quartzite stone finishes
22. Cornice made of steel plates
23. 300X800mm citywall bricks
24. 8+a+8 tempered glass
25. Quartzite stone finishes
26. Flower beds
27. 8+a+8 tempered glass
28. 8+a+8 tempered glass
29. Steel plate
30. 40x80mm heat-saving aluminum windows
31. 8+a+8 tempered glass
32. Cornice made of steel plates
33. 300X800mm citywall bricks
34. Quartzite stone finishes

1. 钢板做檐口（防水防锈处理，深色金属漆饰面）
2. 300mmX800mm城砖（磨砖对缝处理）
3. 新作断桥铝窗
4. 新作紫铜防盗门
5. 青石板石材饰面
6. 花池（黑金砂石材饰面）
7. 钢板做檐口（防水防锈处理，深色金属漆饰面）
8. 板瓦
9. 脊背
10. 钢板做檐口（防水防锈处理，深色金属漆饰面）
11. 300mmx800mm城砖（磨砖对缝处理）
12. 干挂深色铝单板
13. 铜板（防水防锈处理，深色金属漆饰面）
14. 花池（黑金砂石材饰面）
15. 坡道（青石板石材饰面）
16. 黑金砂石材饰面
17. 青石板石材饰面
18. 青石板石材饰面
19. 水池（黑金砂石材饰面）
20. 8+a+8钢化玻璃
21. 青石板石材饰面
22. 钢板做檐口（防水防锈处理，深色金属漆饰面）
23. 300mmx800mm城砖（磨砖对缝处理）
24. 8+a+8钢化玻璃
25. 青石板石材饰面
26. 花池（黑金砂石材饰面）
27. 8+a+8钢化玻璃
28. 8+a+8钢化玻璃
29. 钢板（防水防锈处理，深色金属漆饰面）
30. 40mmX80mm断桥铝窗
31. 8+a+8钢化玻璃
32. 钢板做檐口（防水防锈处理，深色金属漆饰面）
33. 300mmx800mm城砖（磨砖对缝处理）
34. 青石板石材饰面

JIULIHE RESTAURANT
九里河餐厅

When people step into the space, an ecological, vital, lively, lush and active space impression comes into being immediately. The forming of this impression comes from the space in accordance with the environment of wetland park, from the Party A's anticipation of "low-input, high-quality", and even from designers' deep understanding and skilful tactics of "ecological" space. Firstly, in plane layout, apart from accomplishing the dining function, the designer highlights the modern business style of experience, leisure and international sense. Secondly, in the use of materials, a large number of environment-friendly materials are adopted, which not only follows the low-carbon trend, but also is in accordance with "Wetland Park" in spirit. Finally, through the use of rich and bright colors, and the shape of flowers and other plants as well as the presentation of natural texture, characteristics of a vital dining environment are sketched out.

Design Company	设计公司
S-zona Designer Consultant Inc.	无锡上瑞元筑设计制作有限公司
Designer	设计师
Jiayun Feng	冯嘉云
Design Team	参与设计
Ronghua Lu	陆荣华
Project Location	项目地点
Wuxi, Jiangsu	江苏无锡
Project Area	项目面积
2,200 m²	
Main Materials	主要材料
Marble, plating stainless steel, lesco wood, leather with print pattern, wallpaper with print pattern, white figured wood, northeast China ash relief panels, old wood, etc.	大理石、电镀不锈钢、绿可木、皮革打印图案、墙纸打印图案、白影木、水曲柳浮雕板、老木头等

步入其间，生态的、生机的、有生命感的、葱茏的、积极阳光的空间印象顿生。形成这一印象，有对湿地公园大环境的符合，有"低投入、高品质"的甲方预期，更有设计师对"生态"空间的深刻理解与娴熟的手法展现。首先，在平面布局上，在完成餐饮的功能之外，突出了体验型、休闲型、国际感的现代业态气质；其次，在材料使用上，环保型材料的大量使用，既顺应低碳潮流，又与"湿地公园"保持了精神属性的一致；最后，通过丰富、明快的色彩运用、花卉等植物意向的造型，自然肌理的呈现，勾画出富于生机的就餐环境。

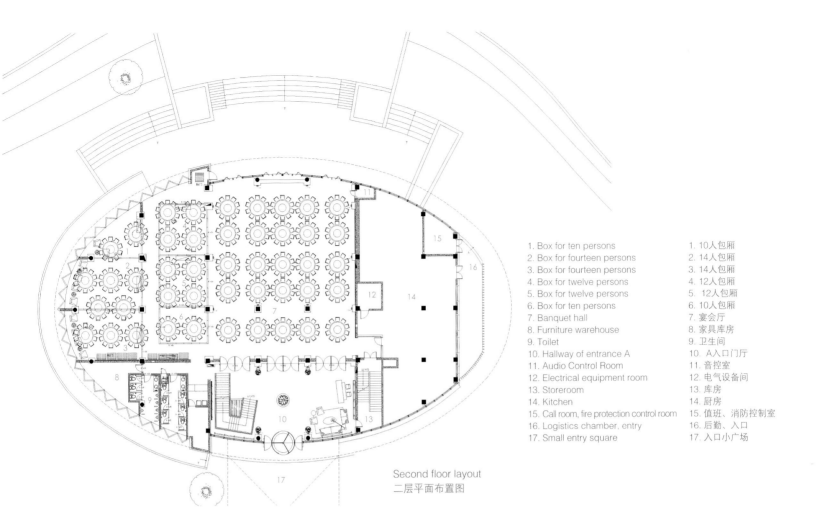

1. Box for ten persons — 1. 10人包厢
2. Box for fourteen persons — 2. 14人包厢
3. Box for fourteen persons — 3. 14人包厢
4. Box for twelve persons — 4. 12人包厢
5. Box for twelve persons — 5. 12人包厢
6. Box for ten persons — 6. 10人包厢
7. Banquet hall — 7. 宴会厅
8. Furniture warehouse — 8. 家具库房
9. Toilet — 9. 卫生间
10. Hallway of entrance A — 10. A入口门厅
11. Audio Control Room — 11. 音控室
12. Electrical equipment room — 12. 电气设备间
13. Storeroom — 13. 库房
14. Kitchen — 14. 厨房
15. Call room, fire protection control room — 15. 值班、消防控制室
16. Logistics chamber, entry — 16. 后勤、入口
17. Small entry square — 17. 入口小广场

Second floor layout
二层平面布置图

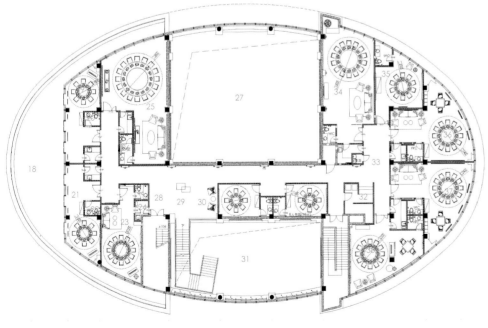

18. Outdoor platform — 18. 户外平台
19. Box D (10 persons) — 19. D包厢(10人)
20. Pantry — 20. 备餐
21. Box C (10 persons) — 21. C包厢(10人)
22. Pantry — 22. 备餐
23. Box A (16 persons) — 23. A包厢(16人)
24. Pantry — 24. 备餐
25. Box E (12 persons) — 25. E包厢(12人)
26. Pantry — 26. 备餐
27. Over the banquet hall — 27. 宴会厅上空
28. Roof garden on the top three floors — 28. 上三层屋顶花园
29. Aisle — 29. 过道
30. Resting area — 30. 休息区
31. Over the hall — 31. 门厅上空
32. Wine reservoir — 32. 酒水库
33. Aisle — 33. 过道
34. Box (14 persons) — 34. 包厢(14人)
35. Box (10 persons) — 35. 包厢(10人)
36. Box (12 persons) — 36. 包厢(12人)
37. Box (12 persons) — 37. 包厢(12人)
38. Box (12 persons) — 38. 包厢(12人)

First floor layout
一层平面布置图

149

BEIJING LONGTAN LAKE PARAMOUNT CHAMBER
龙潭湖九五书院

Chinese palace and historic buildings are the important cultural heritages to be protected, and what they are vividly narrating is not limited to that China is a great nation with a long history, but also includes that it is a civilized nation with glorious culture.

How to transform this profound implication into a design that is in accordance with modern times? In the design, the owner specially stresses "the impression of classical China".

It is a difficult thing for a foreign designer. However, it is just because it is a foreigner that he interprets a kind of new and different beauty. First, considering Chinese historic buildings, the main design has unique stone steps, ceiling and wall decoration. Designers use "matched with some kind of decoration in some part" as the basis of assumption, and then use it as a template to design by transforming the way of presentation and materials, as well as the method of using them.

Design Company 设计公司
SH-RID 上海泷屋装饰设计有限公司
Designers 设计师
Ogawa Norio, Sasaki Chikara 小川训央、佐佐木力
Project Location 项目地点
Beijing 北京
Project Area 项目面积
2,650 m²
Photographing 摄影
Fang Jia 贾方

中国的宫廷以及历史性建筑物等，作为受到保护的重要文化遗产，所有这些在生动叙述着的不仅限于中国是个历史悠久的伟大国家，还包括其是一个拥有光辉文化的文明国度。

如何将这深远的意味，变换成符合现今时代的设计呢？本次设计，业主十分强调"古典中国的印象"。

这对于身为外国人的设计师们是个难题。然而，正因为是外国人，才能演绎出新的、不同的美。首先，就中国的历史性建筑物来说，主要的设计有独特的石阶、天花和墙面装饰。采用"在某个部分配合某种装饰"作为设想的基础，然后以此为范本，通过变换展现方式和素材以及其使用方法等进行设计。

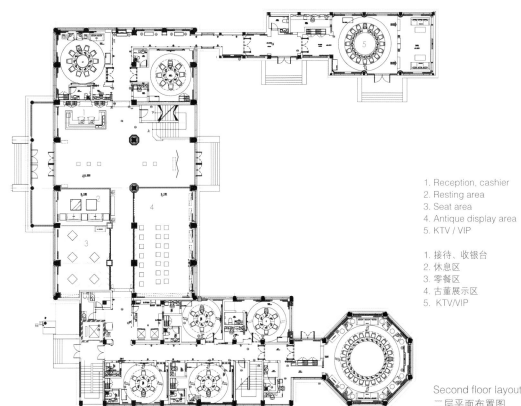

1. Reception, cashier
2. Resting area
3. Seat area
4. Antique display area
5. KTV / VIP

1. 接待、收银台
2. 休息区
3. 零餐区
4. 古董展示区
5. KTV/VIP

Second floor layout
二层平面布置图

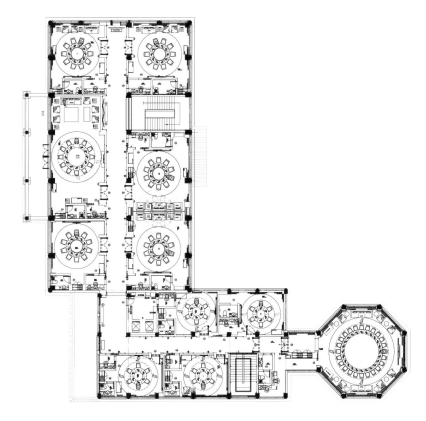

First floor layout
一层平面布置图

　　2650平方米的宽阔会馆，几乎由包厢组成，风格多样的设计使客人能够在不同的包厢内感受到不一样的氛围。比如说，天花上吊有艺术品的房间，强调突出天花的高度，通过改变很久以前就存在的中国传统艺术品的活用方式，来塑造室内氛围的变化。除此之外，通过在天花上贴木材等手法，避免"纯粹古典中式"取向的同时，还要特别注意不能设计成现代中式风格。就设计师的设计初衷而言，希望使客人在"古典"这一设计基调中，同时感受到"当代"特有的时代感。会馆内，也设置了可以摆放古董的展示空间，可以享受到的不仅是店铺的设计，同时还有中国的古典艺术。

　　那是湖畔处若隐似浮的庄严姿容，是朱漆的外表和豪华奢靡的内里，一瞥之间虽是古典之态，但细部满是新中华风韵，很好地表现了前述的设计理念。

IMPRESSION AT JIANGNAN RESTAURANT – WANDA SHOP
印象望江南餐厅万达店

The case is a different attempt of Daohe Design in "Wang Jiang Nan" series restaurant design, which is the combination of New Oriental flavor, southern charm and modern abstract visual experience. The overall design presents a fashionable and concise temperament, and the simple black and white lines have the power straight to the heart.

The contrast of black and white color as well as the scene of specular reflection is endowed with a strong sense of stage. Furniture inside the store gives priority to coffee velvet cloth and marble table, and the wall is mainly shaped with "concavo-convex" lesco wood board and mirror. The strong decorous feeling and the leaping concise lines bring guests a sensory impact, which also reflect the restaurant's brand positioning. The overall color contrast is strong, but the space is full of flow and change.

Design Company 设计公司
Daohe Design Agency 道和设计机构
Designer 设计师
Xiong Gao 高雄
Design Team 设计团队
Yunzong Wu 吴运棕
Project Location 项目地点
Fuzhou, Fujiang 福建福州
Project Area 项目面积
550 m²
Photographer 摄影师
Yuedong Zhou 周跃东
Main Materials 主要材料
Ecological wood, Loulan antique brick, ultra-white glass paint, 5 mm silver mirror, 5 mm black mirror, emulsified glass, black marquina, black stainless steel, etc.
生态木、楼兰仿古砖、超白玻璃烤漆、5mm银镜、5mm黑镜、乳化玻璃、黑白根大理石、黑色不锈钢等

本案是道和设计在"望江南"系列餐厅设计中进行的一次不同尝试，带有新东方情结、江南韵味与现代抽象视觉感受的结合，整体设计呈现时尚、简洁的气质，黑白色相间、简约的线条具有一种直抵人心的力量。

黑白色反差、镜面反射的场景，具有强烈的舞台感。店内家具均以咖啡色绒布和大理石台面为主，墙面造型均以凹凸的绿可板及镜面为主，强烈的厚重感与跳跃简洁的线条给客人带来感观的冲击，也体现了餐厅的品牌定位。整体色彩对比强烈，但空间更富于流动和变化。

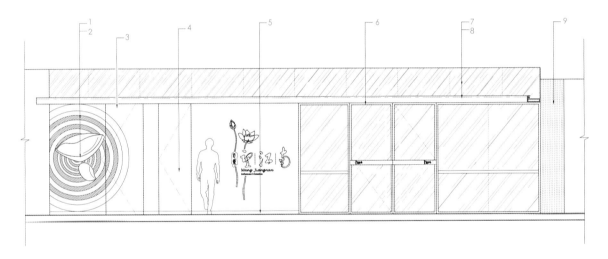

Facade elevation
门面立面图

1. FRP corrugated shape
2. FRP stone shape
3. Invisible doors of kitchen entrance
4. MDF base roasted black paint, fire hydrants door which can be opened
5. Black titanium stainless steel finishes
6. Black titanium stainless steel borders
7. White painted frosted glass finishes
8. Hidden T5 light belt
9. Lesco wood board finishes

1. 玻璃钢波纹造型
2. 玻璃钢石头造型
3. 厨房入口隐形门
4. 中纤板基层烤黑色漆，消防栓门可开启
5. 黑钛不锈钢饰面
6. 黑钛不锈钢边框
7. 白色烤漆玻璃玉砂饰面
8. 暗藏T5灯带
9. 绿可板饰面

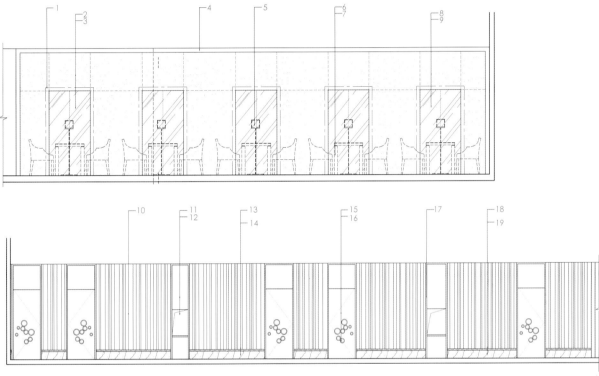

Seat area
散座区

1. Hidden T5 warm color light belt
2. Microlite finishes
3. Ultra-white frosted glass finishes
4. Lesco wood board finishes
5. Floor lamps furnishings
6. Black marquina countertops
7. Black titanium stainless steel table leg
8. Microlite stone finishes
9. Ultra-white frosted glass finishes
10. Lesco wood board finishes
11. Manchurian ash dyed gray finishes
12. Pantry entrance
13. Black titanium stainless steel finishes with built-in T5 lighting belt
14. Black marquina finishes
15. Manchurian ash dyed gray finishes
16. Custom-made stainless steel styling handle
17. Manchurian ash dyed gray finishes
18. Black titanium stainless steel finishes with built-in T5 light belt
19. Black marquina finishes

1. 暗藏T5暖色灯带
2. 微晶石饰面
3. 超白玻玉砂饰面
4. 绿可板饰面
5. 落地灯陈设
6. 黑白根大理石台面
7. 黑钛不锈钢桌脚
8. 微晶石饰面
9. 超白玻玉砂饰面
10. 绿可板饰面（大小型号错拼）
11. 水曲柳染灰色饰面
12. 传菜口
13. 黑钛不锈钢饰面，内藏T5灯带
14. 黑白根大理石饰面
15. 水曲柳染灰色饰面
16. 定制成品不锈钢造型拉手
17. 水曲柳染灰色饰面
18. 黑钛不锈钢饰面，内藏T5灯带
19. 黑白根大理石饰面

1. Kitchen	1. 厨房
2. Reception hall	2. 接待大厅
3. Bufffet area	3. 明档间
4. Hall	4. 大厅
5. Mixed-use area	5. 综合区
6. Box 01	6. 包间01
7. Box 02	7. 包间02
8. Box 03	8. 包间03
9. Box 04	9. 包间04
10. Box 05	10. 包间05
11. Box 06	11. 包间06

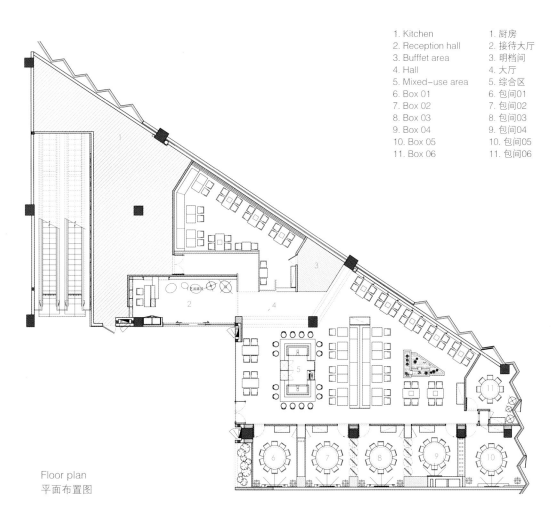

Floor plan
平面布置图

层叠山峦加之跟随的灯光效果与镜面的反射交织出一张具有江南风味的画面，若隐若现的山峦，给人无限遐想。入口大厅的处理是本案的出彩之处，以玻璃钢雕塑或站或坐的各异姿态，加上乳化玻璃发光墙、圆月造型，时尚现代感与东方气质的碰撞，为整个餐厅增添许多各异的色彩。

MANGO THAI RESTAURANT, NINGBO
宁波美泰泰国餐厅

Along with the constantly emerging flavor restaurants of different countries and regions in recent years, Thai restaurants are more and more popular with people. In terms of this case, it is situated on the third floor of the mall and its area is not large. The outside advantage is not obvious, so it is the focus of consideration in design to achieve a good spread by creating the inner advantage of space via unique self recognition. In layout planning, this case anti-conventionally displaces the restaurant entrance to the most remote place of pedestrian flow, so that visitors can perceive the whole dining atmosphere while strolling around. It deliberately produces customers flow in the corridor of restaurant entrance rather than rapid passing, and it separates people flow through a dual-channel shunt after entering the waiting area. Dining area is divided into opposite-sitting area, car seat, casual sitting area and box area according to the population structure.

Design Company 设计公司
Hangzhou Yineiya Architectural Decoration Design Co., Ltd. 杭州意内雅建筑装饰设计有限公司
Designer 设计师
Alex Zhu 朱晓鸣
Design Team 参与设计师
Leon Gao, Huawen Lei, Lulu Zhu 高力勇、雷华文、朱露露
Project Location 项目地点
Ningbo, Zhejiang 浙江宁波
Project Area 项目面积
300 m^2
Main Materials 主要材料
Custom-made antique pattern brick, lime straw mud, old wood, stained oak, MDF carving board, clay tile, outdoor flooring, woven wood, etc.
定制仿古花片砖、石灰稻草泥、老木板、橡木染色、密度板雕刻、陶土瓦、户外地板、编织木等

随着近几年国内不断涌现不同国域、地域的风情餐厅，泰国餐厅也越来越受国人喜爱，就此案来讲，场所位于商场三层，面积并不大，外优势并不明显，而打造空间的内优势，借以独特的自我特征识别，达到良好的传播是设计中考虑的重点。本案在区域划分中反常规地将餐厅入口移到人流交通的最远端，使来访者在迂回的踱步中对整个用餐氛围有所感知，刻意使客户在餐厅门口廊道中有所积流而不是快速分流，进入等待区后，再通过双通道分流。用餐区根据人群结构不同划分了对坐区、卡区、散座区、包厢区等。

　　长条卡区可酌用餐人数快速拼接来改变其接纳量。而泰式礼品区、水吧区、收银区等整合式"中岛"设计，既减少工作人员的数量，又在视觉交叉点上有极佳全景视线，增加了服务的快速便捷性。包厢外走廊的过渡空间划分，使该区域有着远离大厅用餐区视觉错觉感，更加静谧、独立。在设计风格的导入中，考虑其建筑的层高，并结合来访者的年龄层特质，并未一味地将泰式暹罗建筑特质的灿烂辉煌、塔尖翘角等较为异域华丽的元素强加运用。在整体较为质朴、平和、随性的"底色"中略施粉黛，恰当地加入了有泰式民族特色的鲜活图案及色块，进行"矛盾"的破坏，如地毯式定制花砖铺设，阵列的多彩门框，绮丽的家具面料运用增加空间的视觉张力；大厅区、包厢区叠水瓦墙的延续运用加以热带植物的烘托，增加其热带南国的风情。借此打破传统泰式风格的贵族仪式感与单调沉闷。空间既有泰国风情又有再创的现代时尚感与热带气息的轻松用餐环境。

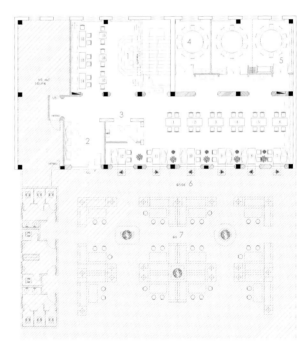

1. Water Bar
2. Rest area
3. Cashier
4. Box 1
5. Box 2
6. Terrace aisle
7. Terrace

1. 水吧台
2. 休息区
3. 收银台
4. 包厢1
5. 包厢2
6. 露台过道
7. 露台

Floor plan
平面布置图

ORIENTAL COURTYARD
东方大院

Colors of different texture are clearly spread in the entire space. Simplified Chinese elements are scattered in every corner, neither indirectly nor straightly, and it is difficult to see through once. There is neither deliberate rendering nor sophisticated decoration in the space and everything seems to be simple yet concise! This kind of understanding and re-creation of the space is from the calmness and joy of the designer's inner heart, and is spread out in the space.

Design Company 设计公司
Daohe Design Agency 道和设计机构
Designer 设计师
Xiong Gao 高雄
Design Team 设计团队
Xianming Gao 高宪铭
Project Location 项目地点
Fuzhou, Fujiang 福建福州
Project Area 项目面积
656 m²
Photographer 摄影师
Yuedong Zhou 周跃东
Main Materials 主要材料
Black gold dragon marble, Equatar Marmara marble, ariston marble, white painted glass, black stainless steel, emulsified glass, etc.
黑金龙大理石、直白纹大理石、雅士白大理石、白色烤漆玻璃、黑色不锈钢、乳化玻璃等

不同质感的色彩被鲜明地铺洒在整个空间内,中式元素经过简化分散在各个角落,没有忸怩,又不是直坦坦的,很难一眼看透;空间中没有刻意的渲染,也没有繁缛的修饰,一切看上去都简简单单又不失凝练!这种对空间的理解与再创作,是来自于设计师内心的淡然和喜悦,并在空间中一阵阵地传递开来。

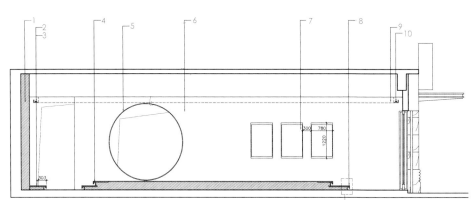

1. Northeast China ash with dark coffee color
2. Hidden T5 warm light belt
3. Equatar Marmara marble finishes
4. Hidden LED warm color light belt
5. 2cm black door pocket with titanium edge
6. 18mm MDF primary aluminum composite panel finishes, putty gray primer white painted finishes
7. 43-inch LED TV, 6cm stainless steel mirror border
8. Silver white dragon marble finishes steps
9. White nitrolacquer finishes solid wood lines
10. Hidden T5 warm color light belt

1. 水曲柳擦深咖色
2. 暗藏T5暖色灯带
3. 直纹白大理石饰面
4. 暗藏LED暖色灯带
5. 两厘米黑钛边宽门套
6. 18mm中纤板基层铝塑板饰面，腻子灰打底白色烤漆饰面
7. LED43英寸液晶电视、6cm镜面不锈钢边框
8. 银白龙大理石饰面踏步(DY-01)
9. 白色硝基漆饰面实木线条
10. 暗藏T5暖色灯带

Entrance corridor facade D construction
入口走廊D立面施工图

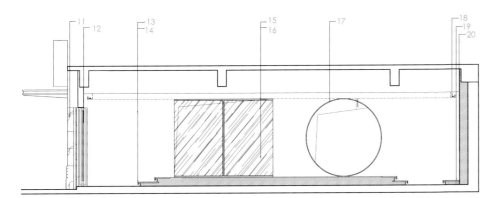

Entrance corridor facade D construction
入口走廊D立面施工图

11. The original building's exterior marble
12. Entrance gate
13. 18mm primary aluminum composite panel MDF veneer, putty gray primer white painted finishes
14. Silvery white marble facade steps
15. Black titanium finishes
16. Two pieces of 5mm ultra-white glass paint glass
17. 2 mm wide door pocket with black titanium edge
18. Hidden T5 warm color light belt
19. Straight veined white marble finishes
20. 18mm MDF primary aluminum composite panel finishes, putty gray primer white painted finishes

11. 原建筑外墙大理石
12. 入口大门（DY-01）
13. 18mm中纤板基层铝塑板饰面腻子，灰打底白色烤漆饰面
14. 银白色大理石饰面踏步（DY-01）
15. 黑钛饰面
16. 两片5mm超白烤漆玻璃（烤漆面对内）
17. 两厘米黑钛边宽门套（DY-01）
18. 暗藏T5暖色灯带
19. 直纹白大理石饰面
20. 墙面18mm中纤板基层铝塑板饰面，腻子灰打底白色烤漆饰面

Floor plan
平面布置图

1. Office
2. Jade warehouse
3. Boutique display area
4. Jade exhibition area
5. Jade exhibition area
6. Calligraphy and painting room
7. VIP room
8. Entrance
9. Bronze lion
10. Bronze lion
11. Furniture display area
12. Agalmatolite boutique display
13. Pantry
14. Business negotiation area
15. Business negotiation area
16. Distribution area
17. Reception room
18. Dressing room
19. Toilet
20. Duty room
21. Control room
22. Tools room
23. Embedded toilet paper holder

1. 办公室
2. 玉石仓库
3. 精品展示区
4. 玉石展示区
5. 翡翠展示区
6. 书画提笔室
7. VIP室
8. 入口
9. 铜狮子
10. 铜狮子
11. 家具展示区
12. 寿山石精品展示
13. 茶水间
14. 商务洽谈
15. 商务洽谈
16. 配电
17. 接待间
18. 更衣间
19. 卫生间
20. 值班室
21. 监控室
22. 工具间
23. 嵌入式手纸箱

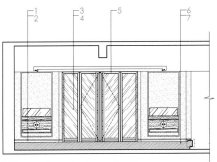

VIP room facade A construction
VIP室A立面施工图

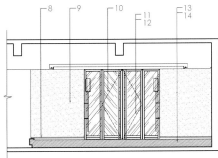

Calligraphy and painting room facade C construction
书画提笔室C立面施工图

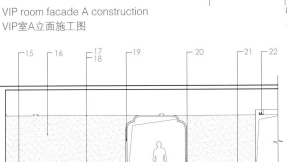

Calligraphy and painting room facade B construction
书画提笔室B立面施工图

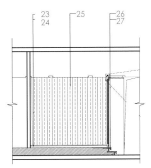

Calligraphy and painting room facade D construction
书画提笔室D立面施工图

1. Cabinet
2. Black titanium baseboard
3. Black titanium closing side door
4. 8mm clear glass finishes
5. Door
6. Cement painted white finishes
7. Bolstered base
8. Black titanium baseboard
9. Cement painted white finishes
10. Door
11. Black titanium side cabinet partition
12. 8mm clear glass finishes
13. Cement painted white finishes
14. Bolstered base
15. Black titanium baseboard
16. Cement painted white finishes
17. Black titanium stainless steel door frame
18. Aisle
19. Door frame
20. Equatar Marmara marble finishes
21. Rubble stone steps
22. Hidden T5 warm color light belt
23. Blue wall covering finishes
24. Black titanium baseboard
25. 60mm organic stick
26. Black titanium finishes door frame
27. Equatar Marmara marble finishes

1. 柜子
2. 黑钛踢脚线
3. 黑钛收边门
4. 8mm清玻饰面
5. 门
6. 水泥漆刷白饰面
7. 垫高基层
8. 黑钛踢脚线
9. 水泥漆刷白饰面
10. 门
11. 黑钛边柜隔断
12. 8mm清玻饰面
13. 水泥漆刷白饰面
14. 垫高基层
15. 黑钛踢脚线
16. 水泥漆刷白饰面
17. 黑钛不锈钢门套
18. 过道
19. 门套
20. 直纹白大理石饰面
21. 毛石踏步
22. 暗藏T5暖色灯带
23. 蓝色墙布饰面
24. 黑钛踢脚线
25. 60mm有机棒
26. 黑钛饰面门套
27. 直纹白大理石饰面

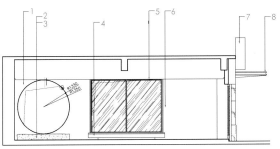

Dry scenic area facade B construction
干景区B立面施工图

Dry scenic area facade C construction
干景区C立面施工图

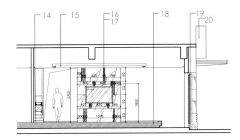

VIP room facade B construction
VIP室B立面施工图

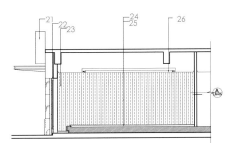

VIP room facade D construction
VIP室D立面施工图

1. 18mm MDF primary aluminum composite panel finishes, putty gray primer white painted finishes
2. 2cm black titanium wide door frame
3. Rubble stone steps
4. Silvery white dragon marble finishes
5. Ceiling
6. 18mm the MDF aluminum composite panel finishes, putty gray primer white painted finishes
7. Advertising light box
8. Original building poncho
9. White latex paint finishes
10. Black titanium closing edge floor-to-ceiling windows
11. White latex paint finishes
12. Black titanium closing edge floor-to-ceiling windows
13. 5cm ceiling closing edge
14. Black titanium baseboard
15. Hidden T5 warm color light belt
16. Equatar Marmara roughened surface marble finishes
17. Equatar Marmara marble finishes
18. Bolstered base
19. Black titanium closing edge floor-to-ceiling windows
20. Advertising light box
21. Advertising light box
22. Black titanium closing edge floor-to-ceiling windows
23. Blue wall covering finishes
24. Blue wall covering finishes
25. Black titanium baseboard
26. Hidden T5 warm color light belt

1. 18mm中纤板基层铝塑板饰面，腻子灰打底白色烤漆饰面
2. 两厘米黑钛边宽门套
3. 毛石踏步
4. 银白龙大理石饰面（DY-01）
5. 吊顶
6. 18mm中纤板基层铝塑板饰面，腻子灰打底白色烤漆饰面
7. 广告灯箱
8. 原建筑雨披
9. 白色乳胶漆饰面
10. 黑钛收边落地窗（DY-01）
11. 白色乳胶漆饰面
12. 黑钛收边落地窗（DY-01）
13. 5厘米吊顶收边
14. 黑钛踢脚线
15. 暗藏T5暖色灯带
16. 直纹白毛面大理石饰面
17. 直纹白大理石饰面
18. 垫高基层
19. 黑钛收边落地窗（DY-01）
20. 广告灯箱
21. 广告灯箱
22. 黑钛收边落地窗（DY-01）
23. 蓝色墙布饰面
24. 蓝色墙布饰面
25. 黑钛踢脚线
26. 暗藏T5暖色灯带

HAO ZI ZAI TEAHOUSE
好自在茶艺馆

Hao Zi Zai teahouse is located inside Fuzhou Ding Guang Temple. The environment of the teahouse is the same as the tea, which is light, refined, restrained rather than pretentious. Due to the geographic advantage of locating in the ancient temple, the teahouse continues the cultural ideology of the ancient "Zen tea": solemn, contemplating beyond all sorrow and happiness, completing the sublimation of humanity in light meaning. To human beings, they can rest half a day in the earthly world and hold a cup of tea in hand to taste the meaning of "tea can purify the heart" and feel the essence of Buddhist pureness brought by tea in the chaos of the world. Isn't it a very great way of leisure?

Design Company 设计公司
Daohe Design Agency 道和设计机构
Designer 设计师
Yushu Guo 郭予书
Design Team 设计团队
Xianming Gao, Yundeng Zhang 高宪铭、张云灯
Project Location 项目地点
Fuzhou, Fujian 福建福州
Project Area 项目面积
260 m²
Photographer 摄影师
Kai Shi, Lingyu Li 施凯、李玲玉
Main Materials 主要材料
Black titanium, square tube, channel steel, linen hard pack, wooden grillwork, wooden wall line, Mongolia black fire burnt plate, white painted glass, etc.
黑钛、方管、槽钢、麻布硬包、木制花格、木制墙边线、蒙古黑火烧板、白色烤漆玻璃等

好自在茶艺馆位于福州定光寺内，茶馆的环境和茶一样，清淡、儒雅、内敛而不张扬。由于坐落在古寺中的地利之便，茶艺馆延续了自古"禅茶一味"的文化思想：静穆，观照，超一切忧喜，亦于清淡隽永之中完成自身人性的升华。于世人而言，在凡尘俗世中偷得半日空闲，持一杯清茶在手，从茶香中品味着"茶可清心"的含义，在纷纷扰扰的世间万象中感受茶带来的佛家清净本质，又何尝不是一种极好的休闲方式？

1. Tea zone
2. Island showcase cabinet
3. Tea table
4. Tea table
5. Cashier
6. Aisle
7. Low showcase cabinet
8. Tea table (8 persons)
9. VIP room
10. Crafts display
11. Male toilet
12. Female toilet
13. Desk
14. Pantry
15. Employees' dressing room and storage room
16. Storage room
17. Box 1
18. Tea table
19. Aisle
20. Box 2
21. Shrine

1. 品茗区
2. 中岛展示
3. 泡茶桌
4. 泡茶桌
5. 收银台
6. 过道
7. 展示矮柜
8. 泡茶桌（8人）
9. VIP包间
10. 工艺品陈设
11. 男卫生间
12. 女卫生间
13. 写字台
14. 茶水间
15. 员工更衣间兼储藏间
16. 储藏间
17. 包间 一
18. 泡茶桌
19. 过道
20. 包间二
21. 佛龛位

Floor plan
平面布置图

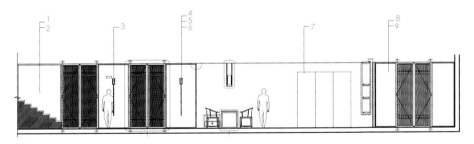

Corridor A elevation
过道A立面图

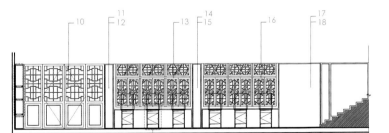

Corridor C elevation
过道C立面图

1. Northeast China ash solid wood line staining
2. White latex paint finishes
3. Art wall lamps
4. Northeast China ash solid wood line staining
5. White latex paint finishes
6. Art wall lamps
7. Finished silk flower screens
8. Northeast China ash solid wood line staining
9. White latex paint finishes
10. Original building doors
11. Northeast China ash solid wood line staining
12. White latex paint finishes
13. Original building windows
14. Northeast China ash solid wood line staining
15. White latex paint finishes
16. Original building windows
17. Northeast China ash solid wood line staining
18. White latex paint finishes

1. 水曲柳实木线染色（黑色硝基漆）
2. 白色水泥漆饰面
3. 艺术壁灯（由设计方指定款式）
4. 水曲柳实木线染色（黑色硝基漆）
5. 白色水泥漆饰面
6. 艺术壁灯（由设计方指定款式）
7. 成品绢花屏风（由设计方指定款式）
8. 水曲柳实木线染色（黑色硝基漆）
9. 白色水泥漆饰面
10. 原建筑门
11. 水曲柳实木线染色（黑色硝基漆）
12. 白色水泥漆饰面
13. 原建筑窗
14. 水曲柳实木线染色（黑色硝基漆）
15. 白色水泥漆饰面
16. 原建筑窗
17. 水曲柳实木线染色（黑色硝基漆）
18. 白色水泥漆饰面

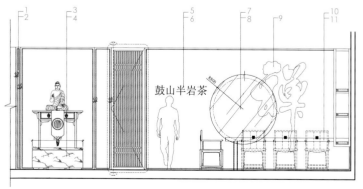

1. Northeast China ash solid wood line staining	1. 水曲柳实木线染色（黑色硝基漆）
2. White latex paint finishes	2. 白色水泥漆饰面
3. Art decorating cabinet	3. 艺术装饰柜（由设计方指定款式）
4. White marble finishes	4. 白色大理石饰面
5. Northeast China ash solid wood line staining	5. 水曲柳实木线染色（黑色硝基漆）
6. PVC advertising characters sprayed roses gold lacquer	6. PVC广告字喷玫瑰金漆
7. Concealed LED light belt	7. 暗装LED灯带
8. White painted glass finishes	8. 白色烤漆玻璃饰面
9. 18mm PCT MDF lettering	9. 18mm密度板刻字（不锈钢管固定）
10. Chinese furniture	10. 中式家具（由设计方指定款式）
11. Log board desktop	11. 原木大板桌面

The hall facade A construction
大厅A立面施工图

茶能使人心静、不乱，有乐趣，但又有节制。本案以精炼的黑白搭配作为主题，简约的直线条，勾勒出复古的空间感，大面积的留白则让空间大气通透。黑白实木的沉静稳重，隐示茶文化有条不紊的文化发展历程；通过现代工艺手法，用黑色硝基漆处理的镀锌管及槽钢、白色烤漆玻璃、茶色玻璃、深色石材等组合，用于展示柜、门、家具；精致的工艺将原本单一的木线条与生硬的墙面结合，贴切主题，提炼精髓且不繁琐，亦为这忙碌喧嚣的世界带来一份静谧与安详……

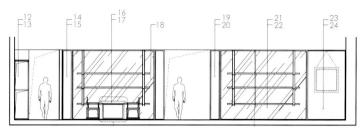

The hall facade B construction
大厅B立面施工图

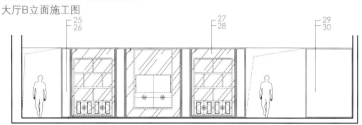

The hall facade C construction
大厅C立面施工图

12. White latex paint finishes
13. Northeast China ash solid wood line staining wardrobe
14. Northeast China ash solid wood line staining
15. White latex paint finishes
16. Channel steel skeleton hanging cabinet
17. Log big board desktop
18. White painted glass finishes
19. Northeast China ash solid wood line staining
20. White latex paint finishes
21. Channel steel skeleton hanging cabinet
22. White latex glass finishes
23. Finished wall picture fixed by hanger
24. White latex paint finishes
25. Northeast China ash solid wood line staining
26. White latex paint finishes
27. Northeast China ash solid wood staining
28. 8 mm white mirror imperial sand neutral glass fixed by glue
29. Manchurian ash solid wood line staining
30. White cement paint finishes

12. 白色水泥漆饰面
13. 衣柜水曲柳实木线染色（黑色硝基漆）
14. 水曲柳实木线染色（黑色硝基漆）
15. 白色水泥漆饰面
16. 槽钢骨架吊柜（固定吊顶）
17. 原木大板桌面
18. 白色烤漆玻璃饰面
19. 水曲柳实木线染色（黑色硝基漆）
20. 白色水泥漆饰面
21. 槽钢骨架吊柜（固定吊顶）
22. 白色烤漆玻璃饰面
23. 成品挂画吊钩固定
24. 白色水泥漆饰面
25. 水曲柳实木线染色（黑色硝基漆）
26. 白色水泥漆饰面
27. 水曲柳实木染色（黑色硝基漆）
28. 8mm白镜御砂中性玻璃胶固定
29. 水曲柳实木线染色（黑色硝基漆）
30. 白色水泥漆饰面

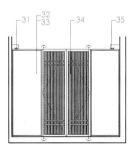

Box 2 facade A construction
包间二 A立面施工图

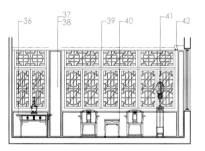

Box 2 facade B construction
包间二 B立面施工图

31. Hidden T5 light belt
32. Northeast China ash solid wood line staining
33. White latex paint finishes
34. Northeast China ash wood grille door
35. Hidden T5 light belt
36. Art decorating cabinet
37. Northeast China ash solid wood staining
38. White latex paint finishes
39. Original building windows
40. Chinese furniture
41. Ornaments
42. Hidden T5 light belt

31. 暗藏T5灯带
32. 水曲柳实木线染色（黑色硝基漆）
33. 白色水泥漆饰面
34. 水曲柳实木格栅门
35. 暗藏T5灯带
36. 艺术装饰柜（由设计方指定款式）
37. 水曲柳实木染色（黑色硝基漆）
38. 白色水泥漆饰面
39. 原建筑窗
40. 中式家具（由设计方指定款式）
41. 装饰物（由设计方指定款式）
42. 暗藏T5灯带

FOUR SEASONS MIN FU ROAST DUCK RESTAURANT

四季民福烤鸭店

The richness of different materials and techniques nowadays makes it become a not too difficult thing to create a top-level restaurant with extreme design, but how to let the design serve for the mass consumers on the basis of civilian consumption really needs the real idea. Four Seasons Min Fu pays more attention to the affinity with civilian consumers in design, but because of the urban cultural background of the city, it should stress the cultural atmosphere of the diet and not lose the sense of the times, which becomes the design focus of the case. The modern Chinese style is the design theme of this case, and the rigorous design rules and visual elements of traditional Chinese style are not too much advocated here, but instead, designers intended to add leisure atmosphere that is close to people. In order to be close to the theme of traditional old Beijing roast duck, the color of space is as close as possible to be simple, and the use of materials is close to nature. The old elmwood recovered from buildings is used in a large area, and retains its natural texture effects after being refurbished, which is matched with the quartzite paving ground to show a rustic texture.

Design Company 设计公司
IN · X   北京屋里门外（IN · X）设计公司

Designer 设计师
Wei Wu   吴为

Project Location 项目地点
Beijing 北京

Project Area 项目面积
840 m²

Main Materials 主要材料
Blue stone board, Sinai pearl, gray linen stone, yellow linen stone, stone, hemp gray, flat iron, distressed elm decorative plates, old elm decorative plates, etc.
青石板、西奈珍珠、灰麻条石、黄麻条石、石材、麻刀灰、扁铁、仿旧榆木装饰板、老榆木装饰板等

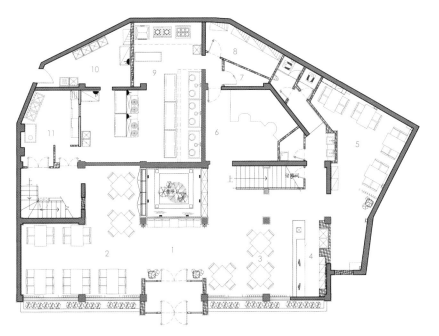

First floor layout
一层平面布置图

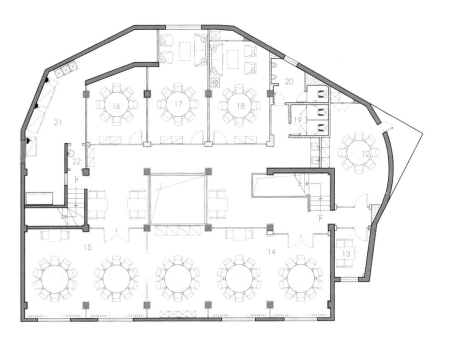

Second floor layout
二层平面布置图

1. Hall	12. VIP room	1. 大厅	12. VIP包间
2. Seat area	13. VIP room	2. 散座区	13. VIP包间
3. Seat area	14. VIP room	3. 散座区	14. VIP包间
4. Cashier	15. VIP room	4. 收银台	15. VIP包间
5. Seat area	16. VIP room	5. 散座区	16. VIP包间
6. Duck room	17. VIP room	6. 鸭房	17. VIP包间
7. Room for prepared duck	18. VIP room	7. 鸭坯间	18. VIP包间
8. Cold storage	19. Female toilet	8. 冷库	19. 女卫
9. Kitchen	20. Male toilet	9. 厨房	20. 男卫
10. Roughing	21. Cold dishes room	10. 粗加工间	21. 凉菜间
11. Dish room	22. Dressing room	11. 洗碗间	22. 更衣间

　　今天的材料种类和技术的丰富，使得用极致的设计打造一个顶级餐厅已经不是太难的事；但如何在平民消费的基础上让设计为大众消费服务，着实需要平实中见思想的真功夫。四季民福在设计上更注重对平民消费的亲和性，但因其所在城市的都市文化背景，又需要对饮食的文化气氛加以强调，并不失时代感，这成为这个案例的设计重点。现代中式是本案的设计主题，对于传统中式厚重严谨的设计规则和视觉元素，这里不做过多提倡，而是意在加入亲民的休闲气息。贴合传统的老北京烤鸭主题，空间的色调氛围尽量贴近淳朴，材料的应用尽可能地贴近自然，大面积使用了从建筑中回收的老榆木，翻新后保留了其天然的肌理效果，配合青石板铺装的地面，呈现出质朴质感。

YILAN COURTYARD HEALTHY CLUB
意兰庭保健会所

Designers adopt the artistic conception of the poem "*Accidental*" by Xu Zhimo to express this case, which apparently shows that what the designers seek is a state of mind, entrusting a kind of emotion, and it is also the spiritual space the public would expect to find. In busy streets, this place is your habitat, creating a comfortable and natural, quiet and relaxed space for you. What the designers seek is just a superficial feeling which blends into Anhui style elements to integrate a most reasonable design space. The infrequent fresh and clean Chinese style with the use of Zen tiles is just like a brush drawing, and the piling up of few materials makes people refreshed. Especially, the use of curtain softens the feel of the entire space. The design of the entire interior space is elegant, quiet, poetic and fun.

Design Company 设计公司
Hefei Xu Jianguo Architectural and Interior Decoration Design Co., Ltd. 合肥许建国建筑室内装饰设计有限公司
Designer 设计师
Jianguo Xu 许建国
Design Team 参与设计师
Tao Chen, Kun Ouyang, Yingya Cheng 陈涛、欧阳坤、程迎亚
Project Location 项目地点
Hefei, Anhui 安徽合肥
Project Area 项目面积
460 m^2
Photographer 摄影师
Hui Wu 吴辉
Main Materials 主要材料
Ancient wood decoration panel, small blue brick, sesame black stone, archaistic plate, etc.
古木纹饰面板、小青砖、芝麻黑石材、仿古板等

设计师借《偶然》(徐志摩)这首诗的意境来表达本案，显然寻求的是一种心境，寄托一种情感，亦是大众所期望寻觅的心灵空间。在冥冥闹市中此处才是你的栖息之所，为你打造舒适自然、安静放松的空间。设计师寻求的恰是蜻蜓点水之情，融入徽派元素整合出最合理的设计空间。少见的清新的中式，带有禅意的瓦片的运用，就像是水墨画一样，而且，少量材料上的堆砌，让人耳目一新。特别是那幔帐的运用，柔化了整个空间的感觉。整个室内空间的设计幽雅、安静、富有诗意与情趣。

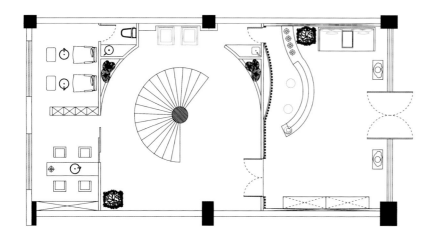

First floor layout
一层平面布置图

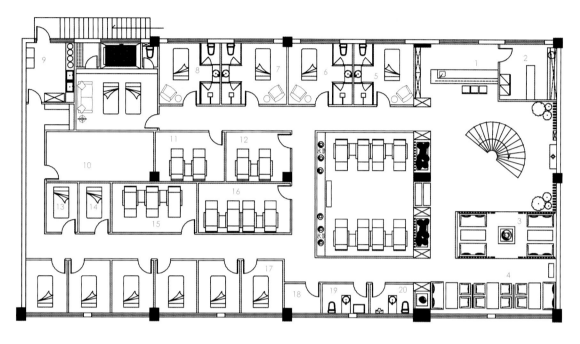

Second floor layout
二层平面布置图

1. Water Bar
2. Workplace
3. Rest area
4. Tea bar area
5. VIP single room
6. VIP single room
7. VIP single room
8. VIP single room
9. Open deployment room
10. Technicians rest room
11. Foot Bath Double
12. Foot Bath Double
13. Single room
14. Single room
15. Foot bath Triple
16. Foot bath Quadruple
17. Single
18. Disinfection
19. Female toilet
20. Male toilet

1. 水吧台
2. 工作间
3. 休息区
4. 茶吧区
5. VIP单间
6. VIP单间
7. VIP单间
8. VIP单间
9. 敞开式调配间
10. 技师休息间
11. 足浴双人间
12. 足浴双人间
13. 单人间
14. 单人间
15. 足浴三人间
16. 足浴四人间
17. 单人间
18. 消毒间
19. 女卫
20. 男卫

TI XIANG YI TEAHOUSE
提香溢茶楼

The design of this case is inspired by the inheritance of traditional essence. Tea culture is originally one of the classical Chinese lifestyles. The establishment of the main theme of Chinese style, through the modern and simple design language to describe, narrows down the distance between such a cultural space filled with fragrance of tea and modern life.

In control of color, the entire space is decorated with sedate warm color, which is matched with the dealing of partial lighting, making the relationship between the space and people closer with the intimate and warm visual experience. The moon gate shape which is commonly used in traditional courtyard design is improved to use in a new way. The waterscape on the ground level reappears the form of the moon gate, but is functionally extended to waterscape design, while handling the partition door of several private rooms extends the concept of moon gate, which presents the original classic style in the form of traditional porcelain vase, bringing out new visual effect.

Design Company 设计公司
IN·X 北京屋里门外（IN·X）设计公司
Designer 设计师
Wei Wu 吴为
Project Location 项目地点
Beijing 北京
Project Area 项目面积
520 m²
Photographing 摄影
Xiangyu Sun 孙翔宇
Main Materials 主要材料
Article lime hemp, quartzite, emulation grass, leather, wood, etc.
条石灰麻、青石板、仿真草、皮、木条等

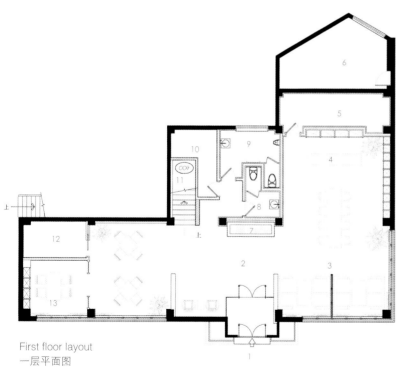

First floor layout
一层平面图

1. Entrance
2. Hall
3. Tea area
4. Bar
5. Operating room
6. Staff room
7. Waterscape
8. Female toilet
9. Male Toilet
10. Storage room
11. Stairwell
12. Storage room
13. VIP room 6

1. 入口
2. 门厅
3. 品茶区
4. 吧台
5. 操作间
6. 员工用房
7. 水景
8. 女卫
9. 男卫
10. 储藏室
11. 楼梯间
12. 储藏室
13. VIP 6

　　本案的设计灵感来源于对传统精髓的继承，茶文化本就是中国经典的生活方式之一，中式风格主调的确立，通过现代简洁的设计语言来描述，将这样一处充满茶香的文化空间，拉近了与现代人生活之间的距离。

　　在对色彩的控制上，整个空间被饰以稳重的暖色调，配合局部光源的处理，以亲切温馨的视觉体验让空间与人之间的关系更加紧密。传统庭院设计中常用的月亮门造型，被加以改进，以新的方式运用。一层的水景再现了月亮门的形式，但从功能上延展为水景的设计，而在几处包房的隔门处理上，则是延展了月亮门的概念，将原本的经典造型，以传统瓷瓶的剪影形式呈现，带来新的视觉效果。

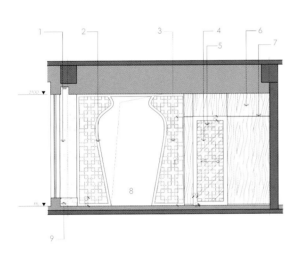
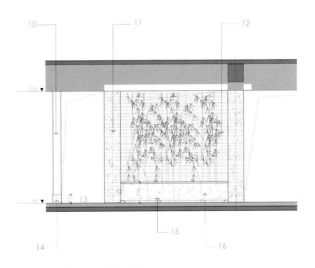

First floor tea area elevation
一层品茶区立面图

First floor hall elevation
一层门厅立面图

1. Rendering
2. Wood finishes
3. Wood finishes
4. Wood finishes
5. Glass
6. Wood finishes
7. 10X10mm groove
8. Access to VIP6
9. Stone
10. Rendering
11. Stone
12. Stone
13. Access to bathroom
14. Stone
15. Hidden Tubing Lamp
16. Stone

1. 批刮灰
2. 木饰面
3. 木饰面
4. 木饰面
5. 玻璃
6. 木饰面
7. 10mmX10mm凹槽
8. 通往VIP6
9. 石材
10. 批刮灰
11. 石材
12. 石材
13. 通往卫生间
14. 石材
15. 暗藏管灯
16. 石材

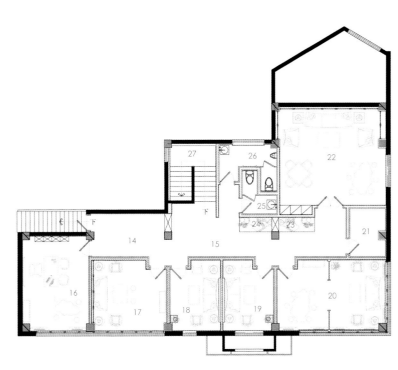

Second floor layout
二层平面图

14. Display area	21. Operating rooms	14. 展示区	21. 操作间
15. Corridor	22. VIP room 1	15. 走廊	22. VIP 1
16. Office	23. Landscape	16. 办公室	23. 景观
17. VIP room 5	24. Landscape	17. VIP 5	24. 景观
18. VIP room 4	25. Female toilet	18. VIP 4	25. 女卫
19. VIP room 3	26. Male toilet	19. VIP 3	26. 男卫
20. VIP room 2	27. Stairwell	20. VIP 2	27. 楼梯间

CAI DIE XUAN IN YANGZHOU
扬州采蝶轩

How to add the fashionable spirit of the times into an architecture with the typical Chinese spirit is the first part of the original intention of designing the project; how to make an international chain restaurant into an individual space with humanity is also a key part; how to form the compatibility of space tone and intensify the identifiability of space is also a part. The final result of multiplex consideration describes the different spatial quality of Cai Die Xuan—a meaningful and exquisite food trip experience full of active fun.

Design Company 设计公司
S-zona Designer Consultant Inc. 无锡上瑞元筑设计制作有限公司

Designer 设计师
Liming Sun 孙黎明

Design Team 参与设计
Hongbo Hu, Xiaoan Xu 胡红波、徐小安

Project Location 项目地点
Yangzhou, Jiangsu 江苏扬州

Project Area 项目面积
1,100 m²

Main Materials 主要材料
Northeast China ash veneer, new castle gray marble, colorful cloud marble, leather, straw wallpaper, silk laminated glass, etc.
水曲柳饰面板、新古堡灰大理石、彩云飞大理石、皮革、草编墙纸、夹绢玻璃等

　　一个典型中式精神的建筑，如何融进风尚的时代精神气质，这是本项目设计初衷的第一部分；如何将一个国际连锁餐饮店打造成具有本埠人文的个性化情境空间，这又是一个重点部分；如何形成空间基调的亲和与强化空间可识别性，这又是一部分。多元的考量最终结果勾兑出扬州彩蝶轩与众不同的空间特质———一道隽永精致充满活性趣味的食之旅体验风景。

典型的中式建筑外立面，通过构成手法获得了均衡与节奏的现代感受，产生了一种"突破深宅大院"的视觉张力，而这种精神在室内空间表现上异曲同工的呼应，灯光透过球形网灯，产生了蝴蝶破茧的自由唤醒。旷达的空间骨架在多种饱满的色彩、多处结构梁架、多种材质及肌理和图形元素中完成了对丰富的演绎，对餐饮气氛的渲染。蝴蝶、琼花等本埠人文诉求则通过当代手法的表现获得了新颖表达。

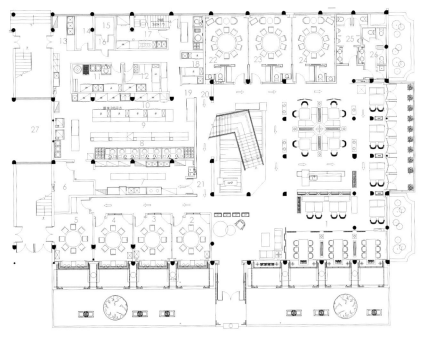

1. Hall, aisle	1. 大厅散座、走道
2. 1C box	2. 1C包间
3. 1C box	3. 1C包间
4. 1D box	4. 1D包间
5. 1E box	5. 1E包间
6. Warehouse	6. 仓库
7. Dish washing room	7. 洗碗间
8. Hot kitchen	8. 热厨
9. Hot dishes assistant	9. 打荷
10. Cutting	10. 配切
11. Cold dishes room	11. 冷菜间
12. Roast room	12. 烧腊间
13. Rough cut	13. 粗加工
14. Locker room	14. 更衣室
15. Freeze room	15. 冷冻
16. Refrigerator	16. 冷藏
17. Snack room	17. 点心间
18. Water room	18. 水吧
19. Pantry	19. 备餐
20. Exit for sending dishes	20. 送菜出口
21. Recycle entrance	21. 回收入口
22. 1B box	22. 1B包厢
23. 1B box	23. 1B包厢
24. 1B box	24. 1B包厢
25. Male toilet	25. 男卫
26. Female toilet	26. 女卫
27. Pavilion	27. 亭

First floor plan
一层平面布置图

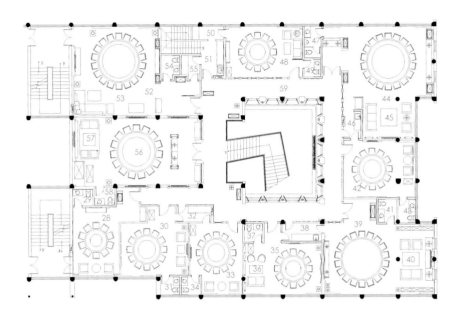

Second floor plan
二层平面布置图

28. 2J box
29. Toilet
30. 2H box
31. Toilet
32. Pantry
33. 2G box
34. Toilet
35. 2F box
36. Rest area
37. Toilet
38. Pantry

39. 2E luxurious box
40. Rest area
41. Toilet
42. 2D box
43. Toilet
44. 2C luxurious box
45. Rest area
46. Pantry
47. Toilet
48. 2B box
49. Toilet

50. Pantry
51. 2M public pantry
52. 2A luxurious box
53. Rest area
54. Toilet
55. Pantry
56. 2K box
57. Rest area
58. Toilet
59. Aisle

28. 2J包厢
29. 卫生间
30. 2H包厢
31. 卫生间
32. 备餐间
33. 2G包厢
34. 卫生间
35. 2F包厢
36. 休息区
37. 卫生间
38. 备餐间

39. 2E豪华包厢
40. 休息区
41. 卫生间
42. 2D包厢
43. 卫生间
44. 2C豪华包厢
45. 休息区
46. 备餐间
47. 卫生间
48. 2B包厢
49. 卫生间

50. 备餐间
51. 2M公共备餐间
52. 2A豪华包厢
53. 休息区
54. 卫生间
55. 备餐间
56. 2K包厢
57. 休息区
58. 卫生间
59. 走道

ORIGINAL MEAL
原膳

"Nature's treasures on Silk Road, outstanding people in eternal city". The case is located in Urumqi, the capital of wide and rich Xinjiang Province in China's western part. The ancient Silk Road brings the pottery and porcelain, silk and lacquer which carry the essence of traditional oriental culture to this Holy Land. At present, Oriental Zen finds here a habitat for soul among glitz and hustle. The designers of this case create Oriental Zen with the dark keynote and outline a space of the rich layering sense with concise and strong straight lines; in a simple and plain manner, a "new trend of old rhyme" is achieved through the virtual and real, bright and dark, simple and complicated dialectical combination.

Design Company 设计公司
Xupin Design Co., Ltd. 叙品设计装饰工程有限公司
Chief Designer 主案设计师
Guoxing Jiang 蒋国兴
Design Team 参与设计
Zhennan Tang, Haiyang Li, Shaoyou Jiang 唐振南、李海洋、蒋少友
Project Location 项目地点
Urumqi, Xinjiang 新疆乌鲁木齐
Project Area 项目面积
5,000 m²
Photographing 摄影
Guoxing Jiang 蒋国兴
Main Materials 主要材料
Black granite, wood veneer, cultured stone, light gray wallpaper, dark wood floors, metal curtain, etc.
黑色花岗岩（荔枝面处理）、木饰面、文化石、浅灰色壁纸、深色木地板、金属帘等

"物华天宝丝绸路，人杰地灵千古城"，本案位于我国西部广阔而又富饶的宝地新疆首府乌鲁木齐。古丝绸之路把承载着东方传统文化精髓的陶瓷、丝绸、漆器带到这片圣土。而今东方禅意在浮华与喧嚣中在这里找到一片灵魂的栖息之地。本案设计师采用深色基调营造东方禅悟。用简洁硬朗的直线条勾勒富有层次感的空间；以朴实的手法，通过虚与实，明与暗，简与繁的辩证结合实现一种"古韵新风"。

设计师并不拘泥于细节的刻画，而是用大面积的面块带过，并恰到好处地用一些中式元素来烘托意境，给人更多的留白空间去品味思索。东方主题的艺术陈设可以淡雅如君子，可以贵气如帝王，可以轻描淡写，可以浓妆淡抹，各具风骚。入口处富有创意的白色同心圆背景墙从视觉上给人强烈的冲击，有种改变地球引力的洒脱；错落有致的流水鸟笼造景，大气古朴的传统宣纸灯具，兴趣盎然的陶瓷漆器制品，宁静致远的白描挂画，这些元素勾勒的景象，乍看若有若无，却让人难以忘怀；超大面积的包厢，配以简洁大方的现代明式家具，意味深远的山水枯木造景，透出淡定的奢华的黑色皮质沙发，整个宏伟包厢简约宁静，让人陶醉，似在诉说男人的情怀。是的，以禅的风韵来诠释室内设计，不求华丽，旨在体现人与自然的沟通，以求为现代人营造一片灵魂的栖息之地。在这里或就餐饮酒，或交友会客，或商务洽谈都从容惬意。这里提倡一种新的生活方式。

现代生活的浮华与喧嚣让我们越来越向往简约宁静的生活，一杯清茶，一抹晚风，都令人陶醉。我们渴望从世俗中解脱，希望从繁杂中抽身，企盼身体与灵魂剥离于凡尘。我们终究是凡人，逃脱不了尘世，但我们可以在这里寻找心灵的慰藉。

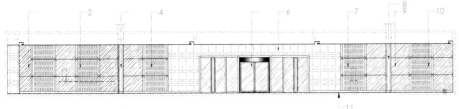

Elevation A in hall
大厅A立面图

Elevation A in hall
大厅A立面图

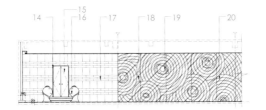

Elevation B in hall
大厅B立面图

1. Glass curtain wall
2. Black metal curtain
3. Black paint
4. Black metal curtain
5. Revolving door
6. Litchi surface stone
7. Black metal curtain
8. Black paint
9. Glass curtain wall
10. Black metal curtain
11. Glossy stone platform
12. Chinese black granite
13. Landscape stone carving
14. Chinese black granite door frame
15. Custom-made wood doors
16. Stone drum guardrail
17. Chinese black granite
18. Landscape sand
19. Landscape stone
20. 12mm tempered glass

1. 玻璃幕墙
2. 黑色金属帘
3. 漆黑色
4. 黑色金属帘
5. 旋转门
6. 荔枝面石材
7. 黑色金属帘
8. 漆黑色
9. 玻璃幕墙
10. 黑色金属帘
11. 光面石材地台
12. 中国黑花岗石（荔枝面处理）
13. 景观石雕（人造仿石）
14. 中国黑花岗石门套（荔枝面处理）
15. 定制木作门
16. 石鼓护栏
17. 中国黑花岗石（荔枝面处理）
18. 景观沙
19. 景观石头（人造仿石）
20. 12mm钢化玻璃

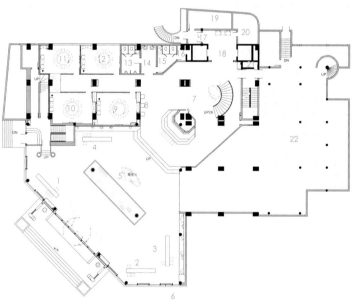

First floor plan
一层平面布置图

Second floor plan
二层平面布置图

1. Reception room 1
2. Reception room 2
3. Reception room 3
4. Reception room 4
5. Sculpture desk
6. Black metal curtain fix
7. Brown wood veneer
8. Decoration desk
9. Box 1
10. Box 2
11. Box 3
12. Box 4
13. Male toilet
14. Wash room
15. Female toilet
16. Cleaning room
17. VIP channel
18. Elevator
19. Distribution room
20. Distribution room
21. Distribution room
22. Kitchen

1. 会客室一
2. 会客室二
3. 会客室三
4. 会客室四
5. 雕塑台
6. 黑色金属帘固定
7. 褐色木饰面
8. 装饰台
9. 包间一
10. 包间二
11. 包间三
12. 包间四
13. 男卫
14. 洗手间
15. 女卫
16. 保洁室
17. VIP通道
18. 电梯间
19. 配电室
20. 配电室
21. 配电室
22. 厨房

23. Dark wood floor
24. Dark wood floor
25. Dark wood floor
26. Dark marble
27. Dark wood floor
28. Dark wood floor
29. Female toilet
30. Dark marble
31. Male toilet
32. Dark marble
33. Dark wood floor
34. Dark wood floor
35. Dark wood floor
36. Dark wood floor
37. Dark wood floor
38. Cold dishes room
39. Cellar
40. Aisle 1
41. Dead trees landscape
42. Landscape
43. Aisle 2
44. Aisle 3

23. 深色木地板
24. 深色木地板
25. 深色木地板
26. 深色大理石
27. 深色木地板
28. 深色木地板
29. 女卫
30. 深色大理石
31. 男卫
32. 深色大理石
33. 深色木地板
34. 深色木地板
35. 深色木地板
36. 深色木地板
37. 深色木地板
38. 凉菜间
39. 酒窖
40. 过道一
41. 枯树景观
42. 景观
43. 过道二
44. 过道三

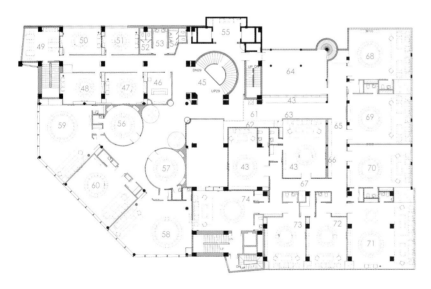

Third floor plan
三层平面布置图

45. Stairwell	60. Luxurious large box 3	45. 楼梯间	60. 豪华大包间三
46. Cashier	61. Aisle 4	46. 收银区	61. 过道四
47. Box 5	62. Landscape	47. 包间五	62. 景观
48. Box 6	63. Dead trees landscape	48. 包间六	63. 枯树景观
49. Scattered units	64. Cold dishes room	49. 散台	64. 凉菜间
50. Box 7	65. Aisle 5	50. 包间七	65. 过道五
51. Box 8	66. Landscape	51. 包间八	66. 景观
52. Female toilet	67. Aisle 6	52. 女卫	67. 过道六
53. Wash room	68. Luxurious box 1	53. 洗手间	68. 豪华包间一
54. Male toilet	69. Luxurious box 2	54. 男卫	69. 豪华包间二
55. Elevator	70. Luxurious box 3	55. 电梯间	70. 豪华包间三
56. Large box 1	71. Luxurious box 4	56. 大包间一	71. 豪华包间四
57. Large box 2	72. Luxurious box 5	57. 大包间二	72. 豪华包间五
58. Luxurious large box 1	73. Luxurious box 6	58. 豪华大包间一	73. 豪华包间六
59. Luxurious large box 2	74. Luxurious box 7	59. 豪华大包间二	74. 豪华包间七

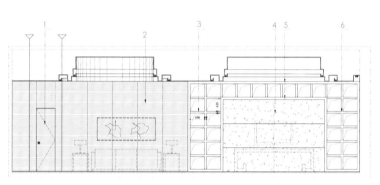

Elevation B in luxurious box
豪华包间B立面图

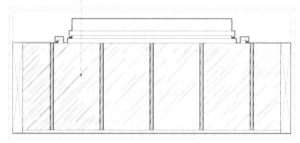

Elevation A in luxurious box
豪华包间A立面图

1. Installation door
2. Custom-made wood finishes
3. Wooden cabinet
4. Dark serpenggiante contraction joints
5. Wood finishes
6. Wooden cabinet
7. Glass curtain wall

1. 安装门
2. 定制木饰面
3. 木作柜
4. 深色木纹石横缝
5. 木饰面
6. 木作柜
7. 玻璃幕墙

JINYU RESTAURANT IN FOSHAN
佛山锦裕食府

The restaurant is located in a high-end community and high-end office area, hiding itself on the second floor, whose design theme is brocade. The layout arrangement of the whole restaurant consists of the hall, Car Seat, semi-open areas and private rooms, so that guests can enjoy more flexible dining options. Large areas of the marble floor with the ancient wood grain in elevator area distribute low-key luxurious atmosphere, and the hollowed phoenix pattern on the wall is reflected via the mirrored stainless steel on the ceiling, which are interwoven and interesting. A strong contrast from the outdoor to indoor makes the conversion of the space form a difference, which allows the guests to slow down for a moment and begin the quality life of the Cantonese-style morning tea.

Design Company	设计公司
Bosidao Design Consultants Limited	博思道设计顾问有限公司
Chief Designer	主案设计师
Zhuyuan Cai	蔡祝源
Design team	参与设计
Hui Shen, Wei Lu, Jianyu Wu	沈卉、陆伟、吴剑雨
Project Location	项目地点
Nanhai, Guangdong	广东南海
Project Area	项目面积
600 m²	
Main Materials	主要材料
Ancient wood grain marble, black stainless steel, oak veneer stain, antique imitation brick, black mirror, etc.	
古木纹大理石、黑色不锈钢、橡木饰面染色、仿古砖、黑镜等	

餐厅位于高端社区及高端写字楼区，藏身于二楼，设计以锦为主题。整体餐厅的格局安排分大厅、卡座、半开放区及包房，让宾客可享受更多更灵活的用餐选择。电梯区大片古木纹大理石地面散发低调奢华气息，墙身的透光镂空凤纹图案通过天花的镜面不锈钢反射效果交织成趣。从室外到室内的强烈对比使空间转换形成落差，让宾客放慢脚步停留半刻，开始广式早茶的优质生活。

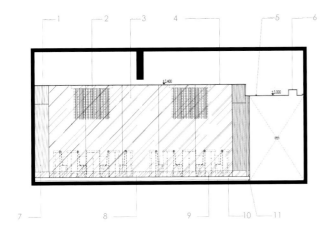

Elevation A in rest area
休闲区A立面图

1. 5mm stainless steel strip
2. Chandelier
3. Art glass
4. Mirror black steel of ceiling bottom
5. 20X20mm groove
6. 300X300mm groove painted black
7. Oak veneer
8. Matte black steel skirting line
9. Hidden spotlights
10. Platform
11. 40X40mm groove

1. 5mm不锈钢条
2. 吊灯
3. 艺术玻璃
4. 天花底镜面黑钢（横向均分6块）
5. 20mmX20mm凹槽
6. 300mmX300mm凹槽刷黑
7. 橡木饰面
8. 哑光黑钢地脚线
9. 暗藏射灯
10. 地台（大理石）
11. 40mmX40mm凹槽

锦，襄邑织文。

——《说文》

古代有"织采为文"、"其价如金"之说，故名为锦。

锦裕食府以回归自然食物原料之本，最大限度地还原天然原料的自然美味，提供独特的味道。强调回归传统手法。吸取各菜系之长，烹调技艺多样善变，用料奇异广博。在烹调上以炒、爆为主，兼有烩、煎、烤，讲究清而不淡，鲜而不俗，嫩而不生，油而不腻，夏秋尚清淡，冬春求浓郁。希望创造返璞归真的菜式和一流的用餐气氛，让客人舒适享受吃的乐趣。

迎面的红酒窖区在暗藏的灯光中映射现代感观，呈现餐厅类而不群的高端定位。其中最具特色的半开放式用餐区，灵感来自高山流水之意，双层玻璃夹胶的简单做法与天花的反射形成天井之意，而画面交织出水墨山水之意境，修身养性寄于品茶餐食内，而品茗食膳于琴瑟之间。

大厅和包房则在整体灰调子下体现锦纹主题。其中包房以中国三大名锦——蜀锦、云锦和床锦为主题。让美食与文化混搭成趣，成为文化传承的亮点。

整个餐厅的照明精心设计，让直接照明、情景照明和装饰照明相互配合成一个整体。灰调子空间呈现独特格调，用心极致的设计细节融汇在特有的东方审美之中，让宾客能在其中细细地尝美食、饮佳酿、品文化。

然美食自成章也。

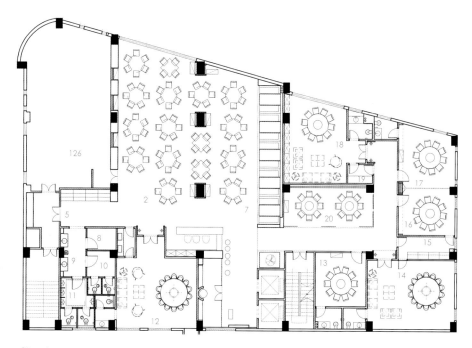

1. Cashier	1. 收银台
2. Hall	2. 大厅（110位）
3. Production area	3. 出品位
4. Seafood pool	4. 海鲜池
5. Recycle area	5. 回收位
6. Kitchen	6. 厨房
7. Box area	7. 厢位
8. Utility room	8. 杂物间
9. Public toilet	9. 公共卫生间1
10. Public toilet 2	10. 公共卫生间2
11. Public toilet 3	11. 公共卫生间3
12. VIP box 1	12. VIP包间1
13. VIP box 2	13. VIP包间2
14. VIP box 3	14. VIP包间3
15. Pantry	15. 备餐间
16. VIP box 4	16. VIP包间4
17. VIP box 5	17. VIP包间5
18. VIP box 6	18. VIP包间6
19. Pantry	19. 备餐间
20. Recreation area	20. 休闲区

Site plan
总平面布置图

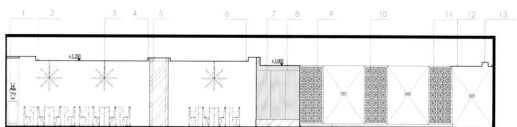

Elevation A in hall
大厅A立面图

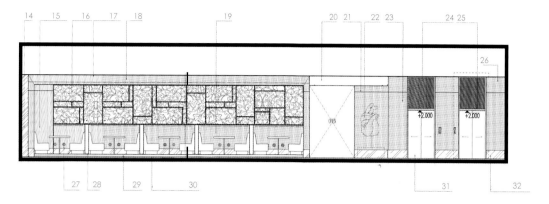

Elevation B in hall
大厅B立面图

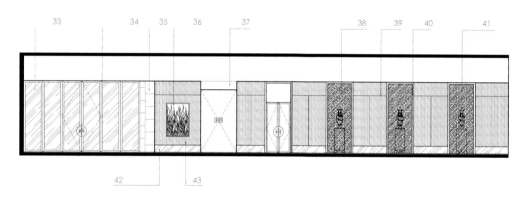

Elevation D in hall
大厅D立面图

1. Art pttery
2. 20X20mm groove
3. Chandelier
4. Painted white ICI
5. Matte black steel
6. 20X 20mm groove
7. Oak veneer
8. 5mm stainless steel strip
9. Matte black steel screen
10. Matte black steel screen
11. Matte black steel screen
12. 20X20mm groove
13. 300X200mm groove painted black
14. Black mirror
15. Hidden yellow T5 lamps
16. Mosaic
17. Hidden yellow T5 lamps
18. Oak veneer
19. Oak veneer
20. White ICI
21. 5mm stainless steel strip
22. Art sculpture
23. Oak veneer
24. Matte black steel with triangle line
25. Hidden yellow lamp
26. Oak veneer
27. Two or three pole socket
28. Hidden spotlights
29. Beige marble
30. 40X40mm groove paste beige marble
31. Elevator door with stainless steel frosted pattern
32. 350mm matte black steel skirting line
33. 15mm steel gray glass
34. 80mm matte black steel closing edge
35. Glossy black steel
36. Art wall picture
37. Painted white ICI
38. Black glass frosted pattern
39. 5mm stainless steel strip
40. Matte black steel
41. Art pottery
42. 350mm matte black steel skirting line
43. Oak veneer

1. 艺术陶艺
2. 20mmX20mm 凹槽
3. 吊灯
4. 刷白色ICI
5. 哑面黑钢
6. 20mmX20mm凹槽
7. 橡木饰面
8. 5mm不锈钢条
9. 哑面黑钢屏风
10. 哑面黑钢屏风
11. 哑面黑钢屏风
12. 20mmX20mm凹槽
13. 300mmX200mm凹槽刷黑
14. 黑镜
15. 暗藏黄色T5灯管
16. 马赛克拼图
17. 暗藏黄色T5灯管
18. 橡木饰面
19. 橡木饰面
20. 白色ICI
21. 5mm不锈钢条
22. 艺术雕塑
23. 橡木饰面
24. 三角线哑面黑钢
25. 暗藏黄色灯管
26. 橡木饰面
27. 二三插座
28. 暗藏射灯
29. 米黄大理石
30. 40mmX40mm凹槽贴米黄大理石
31. 不锈钢磨砂图案电梯门
32. 350mm高哑面黑钢地脚线
33. 15mm钢化灰玻
34. 80mm哑光黑钢收边
35. 光面黑钢
36. 艺术挂画
37. 刷白色ICI
38. 黑玻磨砂图案
39. 5mm不锈钢条
40. 哑面黑钢
41. 艺术陶艺
42. 350mm高哑面黑钢地脚线
43. 橡木饰面

HOKKAIDO TAPPASAKI
北海道铁板烧

The rhythm of the city, the standing tall buildings and the flourishing commercial center, all these make people who live in such an environment for a long period feel suffocated. Designers adopt air, forest, blue sea and sky as well as natural plants as the selected materials of design. Both the real and symbolic forms appeared in the design work, making people full of unlimited reverie. The plants on the wall are matched with the ultraviolet ray lamp, transporting a steady stream of oxygen for the restaurant. The top decoration above the large-scale Tappasaki table in the center is decorated with a large number of logs scattered high and low, symbolizes the forest and ocean together with the auxiliary lighting decoration of the surrounding soft film light blue lighting and also points out the name of "Hokkaido" restaurant. The aisle's 13 meters large-scale gray mirror silhouette light box of land of rivers and lakes makes the design space more concise and concentrated.

Design Company 设计公司
Fuzhou Xinzhu Liu Baoda Design Company 福州新筑设计机构刘宝达室内设计事务所
Designer 设计师
Baoda Liu 刘宝达
Project Location 项目地点
Fuzhou, Fujian 福建福州

都市的节奏，林立的高楼，繁华的商业中心，一切让长时间生活在这样环境中的人窒息。本案设计师将空气、森林、蓝色的海洋和天空，以及自然界植物，作为设计的选取元素，真实与象征的形式出现在设计作品里，让人充满无限的遐想。墙面种植的植物，配合紫外线灯，源源不断地为餐厅输送氧气；中间大型铁板台的顶部装饰，用大量的原木高低错落地装饰，与周围软膜浅蓝光的辅助照明装饰象征了森林与海洋，也为本餐厅的名称"北海道"做一个点题。过道13米的大型水乡灰镜剪影灯箱，使设计空间更加凝练与专注。

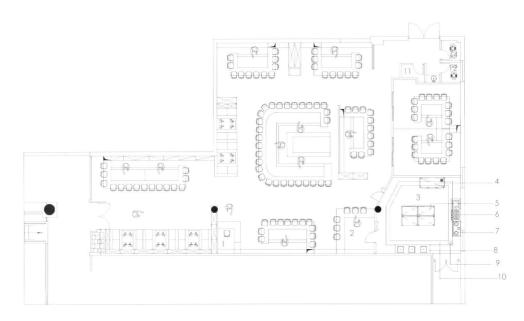

1. Cashier	1. 收银台
2. Sushi station	2. 寿司台
3. Kitchen	3. 厨房
4. Refrigerator with six doors	4. 六门冰箱
5. Operation area	5. 操作台
6. Stoves	6. 六眼灶
7. Large wok	7. 大炒锅
8. Sink	8. 水槽
9. Bottom sinkhole	9. 下排排水口
10. Gutter	10. 排水沟
11. Utility room	11. 杂物间
12. Female toilet	12. 女卫
13. Male toilet	13. 男卫

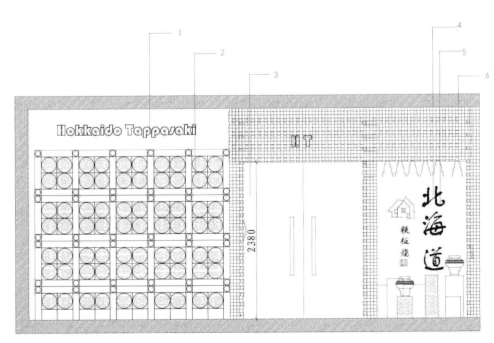

1. Lights in gray mirror
2. Decorative glass
3. 12mm tempered glass door
4. Luminous characters in stainless steel frame
5. Luminous logo in 8mm gray mirror decoration
6. American red oak wood decoration

1. 灰镜内打灯
2. 装饰玻璃
3. 12mm钢化玻璃门
4. 不锈钢框发光字
5. 8mm灰镜装饰内发光标志
6. 美国红橡木实木方装饰

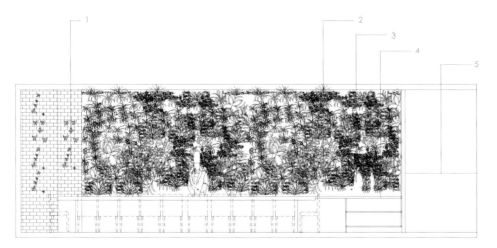

1. Wall decorated by blue brick
2. Plant planted in the ordinary 664 granite
3. Waterproof treatment made in the interior
4. Mongolia black granite
5. Grainy gray marble, with 10mm frosted glass on the top

1. 青砖贴墙面装饰
2. 普通664花岗岩做层板面安装植物
3. 内面做防水处理
4. 蒙古黑花岗岩
5. 木纹灰大理石，上方10mm磨砂玻璃

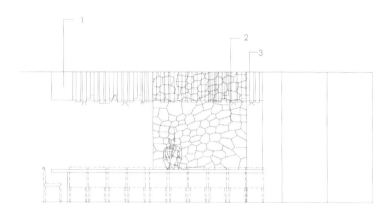
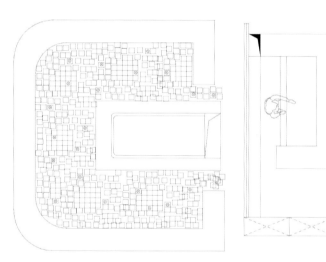

1. Soft film luminous decoration
2. Wood carved surface with metallic paint
3. Square logs decoration

1. 软膜装饰发光
2. 木质刻花面金属漆
3. 原木方装饰

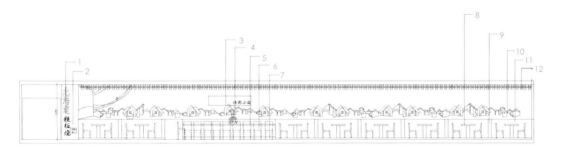
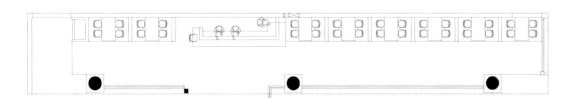

1. Luminous characters in stainless steel frame
2. Bosy gray marble decoration
3. Luminous characters in stainless steel surface
4. Lights in translucent stone
5. Bosy gray marble surface
6. 5mm gray mirror
7. Lights in 10mm tempered glass of the Bosy gray marble base
8. Wooden structure table with Bosy gray marble surface
9. Grainy gray marble with 10mm frosted glass on the top
10. Lights in engraved gray mirror
11. Wall tiling 300X 600mm
12. Square solid wood decorative ceiling

1. 不锈钢框发光字
2. 波斯灰大理石装饰
3. 不锈钢面发光字
4. 透光石内打灯
5. 波斯灰大理石台面
6. 5mm灰镜
7. 波斯灰大理石底座10mm钢化玻璃内打灯
8. 波斯灰大理石台面木结构餐桌面
9. 木纹灰大理石，上方10mm磨砂玻璃
10. 灰镜刻花内打灯
11. 墙面铺贴300mm×600mm
12. 实木方装饰顶

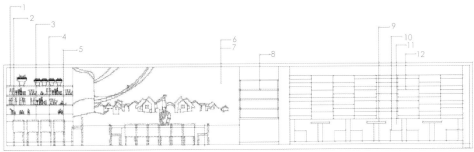

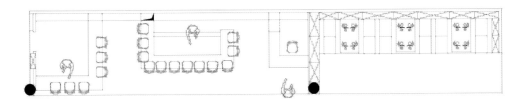

1. American red oak panel veneer
2. Lights in translucent stone
3. Bosy gray marble surface
4. 5mm gray mirror
5. Lights in 10mm tempered glass of the Bosy gray marble base
6. Lights in engraved gray mirror
7. Lights in engraved gray mirror
8. American red oak panel veneer
9. Wooden structure table with Bosy gray marble on the top
10. Custom-made seat
11. Gray mirror wall decoration
12. Colorful painted glass decorated wooden grid

1. 美国红橡木面板贴面
2. 透光石内打灯
3. 波斯灰大理石台面
4. 5mm灰镜
5. 波斯灰大理石底座10mm钢化玻璃内打灯
6. 灰镜刻花内打灯
7. 灰镜刻花内打灯
8. 美国红橡木面板贴面
9. 木结构桌子，面铺波斯灰大理石
10. 定制座椅
11. 灰镜装饰墙面
12. 木质格装饰面贴彩色烤漆玻璃

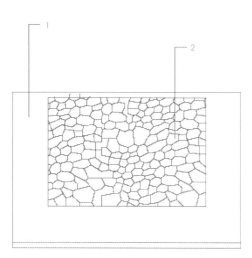

1. Gray mirror wall decoration
2. Wooden carved surface metallic paint

1. 灰镜装饰墙面
2. 木质刻花面金属漆

UNDERSTAND WORLD TEA
观茶天下

The project is located in Huangshan Road, Hefei City, which is the main street with culture of the same strain, with high-level people around it. The designers choose the Huizhou style with the strong tea culture foundation to highlight the characteristics of this case, creating a land of paradise, and the designers try to break through the features of traditional Huizhou architecture to make people enjoy in a relaxed and elegant environment, and slowly feel the essence of Huizhou tea culture. The appearance of this case is orderly arranged with the horse-head wall, which can enhance the impression of Huizhou culture, and make people easily notice this natural pure land. The design ideas are concentrated on the essence of Huizhou tea culture, which is the so-called "Good wine can attract people from all directions and fragrant tea can treat friends coming thousands of miles away", which is exactly the meaning the designers want to express.

Design Company 设计公司
Hefei Xu Jianguo Architectural Interior Decoration Design Co., Ltd. 合肥许建国建筑室内装饰设计有限公司

Chief Designer 主案设计师
Jianguo Xu 许建国

Design Team 参与设计
Tao Chen, Kun Ouyang, Yingya Cheng 陈涛、欧阳坤、程迎亚

Project Location 项目地点
Hefei, Anhui 安徽合肥

Project Area 项目面积
360 m²

Photographer 摄影师
Hui Wu 吴辉

Main Materials 主要材料
Ancient wood grain veneer, small blue brick, sesame black stone, antique imintation plate, etc.
古木纹饰面板、小青砖、芝麻黑石材、仿古板等

　　本案位于合肥市黄山路，是文化一脉相承的主街，周围的人群层次较高。设计师选择具有浓厚茶文化底蕴的徽派风格来彰显本案特点，创造一处世外桃源之地，试图打破传统徽派建筑特点，让人享受一份放松、优雅的环境，细细体会徽州茶文化精髓。本案外观运用马头墙有序排列，可以增强徽文化印象，让人容易注意到这块自然的净土。设计思路主抓徽州茶文化精髓，所谓"酒好可引八方客、茶香可会千里友"，正是设计师所要表达的内质。

　　徽州茶道，讲究以茶立德，以茶陶情，以茶会友，以茶敬宾；设计工作重点是营造茶楼环境、气氛，以求汤清、气清、心清、境雅、器雅、人雅，真正表达博大精深的中华茶文化。

　　本案一楼是茶叶销售区，二楼是品茶区。进入门厅运用书架式隔断，减少外部环境对内部的影响，一楼分为前厅接待区、体验区、休闲景观区、茶叶展示区。茶叶展示区中间有水井相隔开，展区有序地摆放着茶产品，展区四周循环通道，方便顾客流动与选取。一楼景观区有古琴、书卷架、假山水景，让人感受一份平静、朴素、平和、自然的空间氛围。设计师把人造天井运用在本案中，其间的假山水景，巧妙地连接一二两层楼，一楼可以看到人造天井，异常通透，采光效果好，二楼顾客可以围绕天井欣赏一楼布景，鹤与流水的造景相映成趣，给人一种回归自然与纯朴的感觉。二楼饮茶区分服务区、休闲区、包厢区、书画区、卧榻、功能齐全，以满足不同客人的需求。另外还设立冷藏储茶区，将客人所购买的茶叶储藏，方便顾客待客之需。徽派建筑讲究四水归堂，上有天井，下有水景，设计师有意将室内一二层景观相互渗透，在空间中层层相互套接，每一处好似各自独立，却又能融合成一个整体。

　　设计师偶然在家具厂发现的20世纪留下来的废弃的旧桌腿，收购再利用改造成本案的楼梯扶手，给废弃的旧物带来了新的生命，意义非同寻常，新改造的楼梯扶手具有一种仿古的韵味，是本案原始、回归、自然的体现，也表达了设计师的从容和自然，营造一种论茶论道的环境，是种大气的设计手法，从而创造一个为成功人士畅饮通杯没有压力的独特的品茶环境。

　　完成后的作品风格主调明确，与以往的茶楼概念有所不同，将传统装饰元素的经典之处，提炼并演变成为新的设计符号，在二楼品茶区运用了细腻并充满文化气息的细节装饰。本案通过现代简洁的设计语言来描述，将这样一处充满茶香的文化空间，拉近了与现代生活之间的距离。空间中的流动性、透明性、开放性以及互融性，充分体现了设计师与整个空间的艺术理念：即使身在繁杂的大都市，我们依旧能够创造一份纯净的天空。

　　在色彩控制上，整个空间以稳重的暖色调，配合局部光源的处理，以亲切温馨的视觉体验让空间与人之间的关系更加紧密。很多家具运用了原色，原色系意在表现根本、本性、自然的特征，茶的无形之香，使品者反观自己的本性——真、善、美。设计师寓意唤醒茶性和人性的真理。

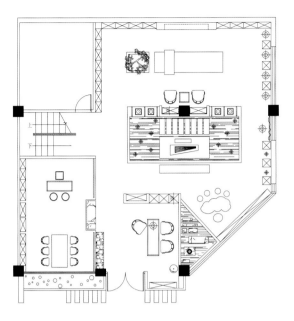

First floor plan
一层平面布置图

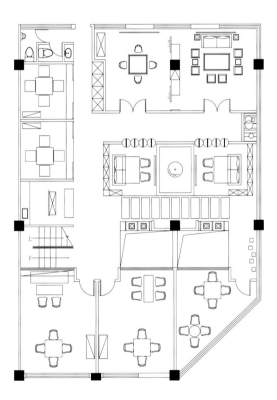

Second floor plan
二层平面布置图

POMEGRANATE BLOSSOM HALL
榴花溪堂

"The benevolent, stable like a mountain, is hard to move by fame while the wise, flowing as a stream, never dies of good thoughts", which is almost known by all the Chinese people. This phrase comes from the *Analects*, and the original sentence of Confucius was: "The wise, flowing as a stream, never dies of good thoughts, while the benevolent, stable like a mountain, is hard to move by fame. The wise are active, and the benevolent are quiet. The wise are happy, while the benevolent have long life." It means that the benevolent are calm and stable, and are not shaken by external things like the mountain. In Confucian view, the natural world should live in harmony. As a product of nature, man and nature are one. In ancient times and ancient fad, people always kept an awe and respect for nature in heart. What's more, having dialogues with nature, keeping harmonic relationship with nature, comparing people themselves to nature fulfill favorable climatic, geographical, human conditions as well as nature unity, which is an otherworldly style, a pure state of mind and a pursuit of sustaining a peaceful world.

Design Company 设计公司
Beijing Libeiya Architecture Engineering Co., Ltd.　北京丽贝亚建筑工程有限公司
Designer 设计师
Aicheng Qiu　邱爱成
Project Location 项目地点
Xian, Shanxi　陕西西安
Project Area 项目面积
2,857 m²
Main Materials 主要材料
Elm, gold brick, cultured stone, wallpaper, etc.　榆木、金砖、文化石、壁纸等

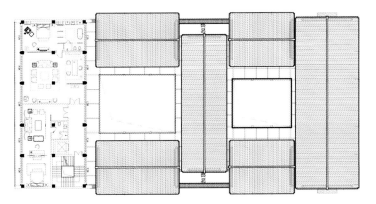

Second floor plan
二层平面图

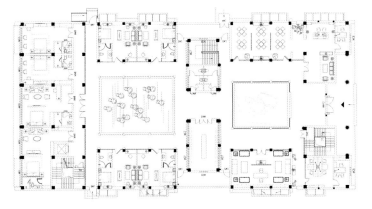

First floor plan
一层平面图

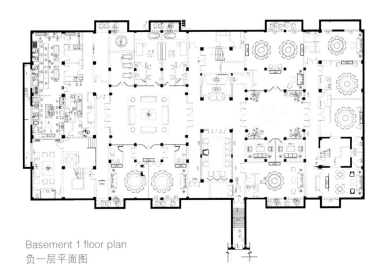

Basement 1 floor plan
负一层平面图

"仁者乐山，智者乐水"，这几乎是所有中国人都耳熟能详的一句话。这句话出自《论语》，孔子当时的原话是这样说的："智者乐水，仁者乐山。智者动，仁者静。智者乐，仁者寿。"其意思是说，仁爱之人像山一样平静，一样稳定，不为外在的事物所动摇。在儒家看来，自然万物应该和谐共处。作为自然的产物，人和自然是一体的。古的时代古的风尚，对于大自然的敬畏和崇敬激荡于古人的胸中，与大自然对话，与大自然相谐，与大自然作比，实现天时地利人和、天人合一，是一种超脱的时尚，是一个洁身自好的境界，甚至是修身治国平天下的追求。

PIN YI SALON
品奕造型

Pin Yi Salon is an integrated shop containing beauty and hairdressing. It's located in the bustling residential area and forms a distinct contrast with various surrounding beauty salons with a gentle and high posture. The pursuit of natural sense and fully relaxed environment is the inspiration of design, but the most direct expression comes from the materials; natural stone, straw wallpaper and other natural materials which are from the nature make this space more intimate and highlight the unique aspect. The open hairdressing space presents a stylish and warm modern style, but the relatively private beauty part combines structural characteristics of the original space to create a modern Chinese style. These two spaces are jointed together by the material and color coordination.

Design Company 设计公司
IN·X 北京屋里门外（IN·X）设计公司
Designer 设计师
Wei Wu 吴为
Project Location 项目地点
Beijing 北京
Project Area 项目面积
354 m²
Main Materials 主要材料
Rock wool board, wooden travertine, straw wallpaper, stone mosaic, laminate flooring, oak veneer, brushed stainless steel, wood decorative panels, clear mirror, etc.
岩石毛板、木纹洞石、草编壁纸、石材马赛克、复合木地板、橡木饰面板、拉丝不锈钢、实木装饰板、清镜等

　　品奕是一间包含美容与美发功能的综合店，地处繁华的居民区，以温婉高调的姿态，与周遭的各色美容美发店形成了鲜明的对比。追求天然感与令人全身心放松的环境，是设计的灵感所在，而最直接的表达方式来自于材质，天然毛石、草编壁纸等源自自然的材料让这一空间更加亲切，也凸显出与众不同的一面。开放的美发空间呈现出时尚温馨的现代风格，相对私密的美容部分则结合了原有空间的结构特色，打造成现代中式风格，两处空间通过材料与色调协调衔接在一起。

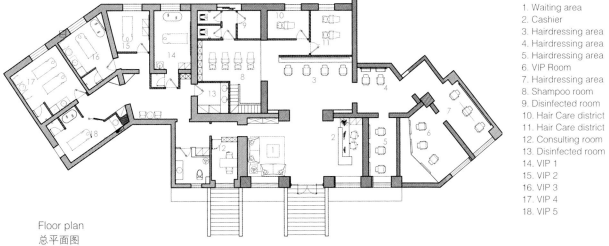

Floor plan
总平面图

1. Waiting area — 1. 等候区
2. Cashier — 2. 收银台
3. Hairdressing area — 3. 美发区
4. Hairdressing area — 4. 美发区
5. Hairdressing area — 5. 美发区
6. VIP Room — 6. VIP室
7. Hairdressing area — 7. 美发区
8. Shampoo room — 8. 洗头室
9. Disinfected room — 9. 消毒间
10. Hair Care district — 10. 头发护理区
11. Hair Care district — 11. 头发护理区
12. Consulting room — 12. 咨询室
13. Disinfected room — 13. 消毒间
14. VIP 1 — 14. VIP 1
15. VIP 2 — 15. VIP 2
16. VIP 3 — 16. VIP 3
17. VIP 4 — 17. VIP 4
18. VIP 5 — 18. VIP 5

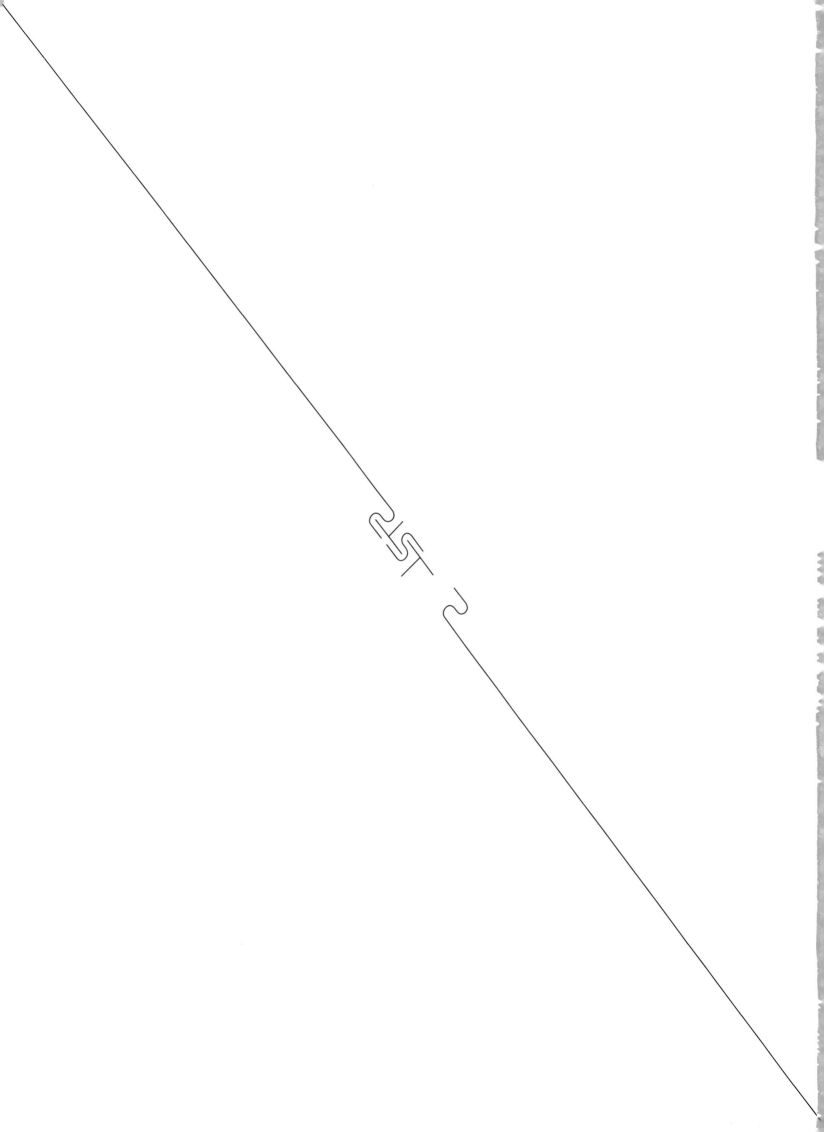